Teleoperation: Numerical Simulation
and Experimental Validation

EURO

COURSES

A series devoted to the publication of courses and educational seminars organized by the Joint Research Centre Ispra, as part of its education and training program.
Published for the Commission of the European Communities, Directorate-General Telecommunications, Information Industries and Innovation, Scientific and Technical Communications Service.

The EUROCOURSES consist of the following subseries:

- Advanced Scientific Techniques
- Chemical and Environmental Science
- Energy Systems and Technology
- Environmental Impact Assessment
- Health Physics and Radiation Protection
- Computer and Information Science
- Mechanical and Materials Science
- Nuclear Science and Technology
- Reliability and Risk Analysis
- Remote Sensing
- Technological Innovation

COMPUTER AND INFORMATION SCIENCE

Volume 4

Teleoperation: Numerical Simulation and Experimental Validation

Edited by

Marc C. Bequet
Commission of the European Communities,
Joint Research Centre,
Institute for Systems Engineering and Informatics,
Ispra, Italy

Springer-Science+Business Media, B.V.

Based on the lectures given during the Eurocourse on
Teleoperation: Numerical Simulation and Experimental Validation
held at the Joint Research Centre, Ispra, Italy, November 18–22, 1991

Library of Congress Cataloging-in-Publication Data

Teleoperation : numerical simulation and experimental validation / edited by Marc C. Becquet.
p. cm. -- (Eurocourses. Computer and information science ; v. 4)
"Based on the lectures given during the Eurocourse on 'Teleoperation: numerical simulation and experimental validation,' held at the Joint Research Centre Ispra, Italy, 18-22 November 1991."
"Published in cooperation with NATO Scientific Affairs Division."
ISBN 0-7923-1584-7 (alk. paper)
1. Manipulators (Mechanism) 2. Remote control. 3. Robotics.
I. Becquet, Marc C., 1960- . II. Commission of the European Communities. Joint Research Centre. Ispra Establishment.
III. Series: Euro courses. Computer and information science ; v. 4.
TJ211.T443 1992
620'.46--dc20 91-46137
CIP

ISBN 978-94-010-5170-5 ISBN 978-94-011-2648-9 (eBook)
DOI 10.1007/978-94-011-2648-9

Publication arrangements by
Commission of the European Communities
Directorate-General Telecommunications, Information Industries and Innovation,
Scientific and Technical Communication Unit, Luxembourg

EUR 14033

Published by Kluwer Academic Publishers,
P.O. Box 17, 3300 AA Dordrecht, The Netherlands.

Kluwer Academic Publishers incorporates the publishing programmes of
D. Reidel, Martinus Nijhoff, Dr W. Junk and MTP Press.

Sold and distributed in the U.S.A. and Canada
by Kluwer Academic Publishers,
101 Philip Drive, Norwell, MA 02061, U.S.A.

In all other countries, sold and distributed
by Kluwer Academic Publishers Group,
P.O. Box 322, 3300 AH Dordrecht, The Netherlands.

TABLE OF CONTENTS

Preface

The solution to today's demand for safety in Industrial Processes and Plants involves more and more the use of teleoperated devices to avoid human exposure to hazards. Such hazards occur during operation and maintenance, in normal or accidental conditions, and during decommissioning at the end of the life time of installations.

The year 1989 has represented, for Teleoperation, an important landmark with the closure of the first commercial nuclear power stations which have reached the end of their life time.

At the same time, a growing concern has generated support for the development of teleoperated devices which could be used in "disordered environments". For example, these could be a Euro-T.M.I. or at worst, a Euro-Tchernobyl, where routinely-employed remote handling equipment could not cope.

Progress in the Controlled Thermonuclear Fusion programmes demonstrates the importance of Teleoperation in the assembly phase of future reactors as well as in the maintenance and decommissioning phases. The design of Fusion Reactors is probably the first example to be driven by Remote Handling concerns.

Non-nuclear hazardous areas, like the chemical industries or the Bio and Genetic engineering industries, also need Teleoperation.

This course is intended to present the fundamentals and the advances in simulation and experimentation in the field of Teleoperation. It is devoted to the presentation of the state of the art in both simulation and experimentation.

Topics like single-master multi-slave manipulator systems and cellular robotic systems, kinematic calibration, motion planning and path finding for autonomous mobile robots are described with the accent on the simulation aspect.

Technological developments for Position Sensing for Advanced Teleoperation in a nuclear environment are stressed with the key issue of sensor hardening for radiation tolerance.

Experimental activities from the Remote Handling Group of the JET Fusion device and of the CERN accelerator give major indications on the validity of simulation and guidelines for items to be improved like the Control Station and Vision in 3D. Decommissioning actions are presented by the BR3 Reactor Dismantling Pilot Project in CEN MOL.

As a perspective of the future, the R&D programmes for Remote Handling for the future Fusion Reactors like NET and ITER, which will be the biggest clients for such products, are described.

From the content of what follows in the next sections, the reader will understand that tomorrow's technology for the safe operation of Hazardous Processes and Plants is produced today all over the world and, in particular, with European partners in a leading position.

Dr. R. W. Witty

SOME KEY ISSUES IN REMOTE HANDLING

Marc BECQUET
Institute for Systems Engineering and Informatics, TP 210
Commission of the european Communities,
Joint Research Center, I-21020, ISPRA, ITALY

Abstract

This paper intend to present some key issues in the future of Teleoperation as it could be felt from the existing technology and the end user demand.
Various aspects are considered like *Mechanical Design, Actuators, Control System* and *Sensors*. These are directly related to manipulators.
Other aspects, concerning a higher layer in the overall architecture of the teleoperated system are also presented: *Viewing and Vision System, Control Station, Design for Maintainability* and *Reliability*.
It is more a personal list of things to think on than an extensive review of existing problems. The reader may find other items to add, depending ones problem or application.

1- Mechanical design

Remote Handling system have appeared with the first experiments in the nuclear field.
In that time, no reference design was available and most of the development has started from scratch. The mechanical design was inspired by aircraft and weapon industries.
In the early sixties, the robotic technology was emerging with Unimate, a company created by J.Engelberger, who is now considered a pioneer in robotic. The interesting fact is that remote handling technology and robotic technology have continued to live mostly ignoring each other in the last twenty years. It is only very recently that common benefit have been identified from the potential collaboration between these two technologies.
In the mechanical design aspects, this benefit is probably the most visible one.
Industrial robots are characterized by good stiffness, large payload and good working envelope. There are also some drawback like lower ratio payload/ weight compared with master slave arm resulting from RH developments.
The gap between these two technologies should be closed in near future with the extensive use of new material like honeycomb material or carbon fiber reinforced plastics. The question of radiation tolerance for such

M. C. Becquet (ed.), Teleoperation: Numerical Simulation and Experimental Validation, 1–5.

materials must be carefully examined to avoid to short life time of structural components.

2- Actuators

In the field of actuation, things have moved from direct mechanical link to allow manual propulsion from safe zone to contaminated zone, to the ultimate electrical direct drive motor or piezo electric micro actuators.
Today, actuators are showing outstanding efficiency figures with compactness. Low speed and high torque capability are available.
This last point was one of the famous drawback of classical DC or AC electric motors. It was the only reason to interpose a reduction gearbox with an increased weight and source of other mechanical failures. Of course, when you eliminate a gearbox is its braking capability are lost if the gearbox was designed to be irreversible. With a direct drive motor you will need in most cases an electrical brake or clutch.
Amplifiers to feed current in electrical motors have also experienced strong improvements from valves in the early fifties to the popular PWM servo amplifiers now available from the shelf. higher switching frequency in transistor based amplifier board reduces dissipation, electrical interferences and noise.
Hydraulic actuators have always show very good compactness with high forces or torques. The sensitive component is still the servo valve that control the admission of hydraulic power in the cylinder. Problem of fluid contamination (particles, bacteria's..) is still present today. Another problem with hydraulics is the leakage of the hydraulic fluid which may causes problem of cleanability. Water is in this case a good substitute to oil.
Modern control technique - see next section- allows to compensate friction effect always present in hydraulic systems and make possible force reflexion: you may back drive an hydraulic piston.

3- Control system

Control theory has found its first theoretical nucleus with modern aircraft manufacturing just before WWII with the invention of the autopilot using the gyroscope compass. It has then moved from control of airplanes to domestic appliance as well as computer technology, electronic chip design etc..
Control systems in remote handling were introduced when mechanical links from master system to slave system have disappeared. Early applications were limited to PID control of the slave arm with respect to the reference position given by the master arm. The next step was force reflexion to give at the operator on the master side the perception of the force exerted by the slave arm. Recent progress in control systems have made possible the elimination of friction effects which degrades strongly the force reflexion capability and above all the sensitivity in force perception. This is mostly done by interposing force and torque sensors at the very extreme of the manipulator structure: the end effector. It could also be done by measuring directly the force or torque inside each joint. These informations can be

retrofitted in the control system for force reflexion but also for dynamic control and hybrid control mixing trajectory and force constraint.
Operator may constrain the end effector to move along a line, onto a plane or forbidden penetration in selected sub space by artificial force reflexion: you may create artificial obstacles for safety reason.
Among the future trends, adaptive control systems will provide the capability of self tuning to overcome drift of mechanical and electrical characteristics.
Neural networks techniques in the field of control systems have already shown interesting results in navigation of mobile robot, control of dual arm, as well as tools for incorporation of both vision, touch and other senses in control loops. Moreover these technique have already been used to improve model accuracy to solve the problem of kinematic model accuracy.
These are the most promising tools of next generation of controllers considering in particular the use of Transputers as processing units.

4- Sensors

Sensing in remote handling is almost what makes the all thing possible.
As in robotics, the perception of the environment and of the configuration of the remote handling system in the workspace makes the mission control feasible. Various type of sensors can be considered: proprioceptive sensor (attached on the structure of the manipulator) and exteroceptive sensor (attached to the global coordinate frame of the workspace). In todays technology, both are used in combination to control the manipulator status - configuration, velocity, acceleration, forces and torques in the joints and at the end effector- and the environment status - scene analysis, object recognition, collision avoidance.
One of the key issue in hostile environment with sensors is reliability and accuracy of data output. In extremely radioactive environment dose rates are to be expected between 0.1 and 10 Mrad/h. Integrated dose of 1 to 10 Grad represent mission times between 100 to 1000h. it means that sensitive parts of sensor must be easily changed and requires design for maintainability.

5- Viewing and Vision System

Viewing and Vision systems are used for years. These systems go from fixed network of cameras with adjustable view point using periscope and prism like IVIS in JET, to mobile telescopic systems able to bring a tele camera over a wall like private eye in CERN. Todays technology makes possible the use of stereo vision with the capability to give not only depth perception but also geometrical position information. Advanced image processing combined with laser scanning for range images are available technologies.
In this field of Viewing and Vision, the ultimate are the stereo vision helmet that give to the operator the impression to be on the spot: telepresence. These products have to mature but will be with no doubt the key component of the future Remote Handling Control stations.

6- Control Station

The control station is a collection of various functionality with the aim to help the operator to perform the mission. among the various functionality, the basic one is the servo control of the remote handling device in use during the mission. Then the operator needs in most case the help of a viewing or vision system. For complex situations, graphic simulation functionality offers a redundant information on the situation of the RH system in the workspace which increases the overall reliability and safety of the mission. Graphic stations are now going closer and closer to real time 3D model handling and are able to mix TV images with CAD ones. The major problem in the future will be the interfacing of these various functionality in order to concentrate the operators attention on the critical points of the mission.

7- Design for Maintainability

As design for assembly is today a strong concern in industrial robotic to enlarge the spectrum of applications, design for Maintainability of a plant is an increasing concern in the design of new plant or the improvement of old ones when remote handling has to be applied.
Some simple facts like avoiding stairs and replacing them by lifts make much easier the design of a mobile robot for maintenance in a NPP or chemical plant.
In the same order of idea, provision for power plugs, high band communication network, viewing systems and accurate 3D Computer Aided Design model of the plant make a large difference in Remote Handling Mission feasibility and Reliability.
In the same order of idea, design for maintainability and cleanability also applies for robotic and remote handling devices.

8- Reliability

Reliability is certainly one of the major concern for RH when one intends to perform maintenance and intervention tasks in hostile environment. Constraints in hostile environment require different way of using and programming of robots when compared with industrial application.
To take maximum benefit of robot technology, the level of *Confidence* in the robotics tool must be much higher than in classical production world. To increase this level of confidence, application of *Reliability Engineering* in combination with strong knowledge of robot technology leads to such an objective.
Different tools in different steps must be used to improve reliability of RH and robotics systems.
The first one is the Analysis of reliability of robotics and in remote handling systems in general to identify failure modes, effects on the system, sensitive components and needs of redundancy and backup systems. Tools as the Failure Modes, Effects and Criticality Analysis are used as well as the Fault Tree Analysis.
The second one deals with design criteria for new robot systems or improvement of existing one using reliability and safety driven design

concepts. Such concepts are applicable on mechanical design, electrical design and electronic design including the computer controller of the robot. The last aspect is the control in real time of availability of functions, safety level as well as failure detection in the various subsystems composing a robot device. Techniques of supervision by use of safety check subroutines are now implemented in the latest generation of control systems.

Conclusion

The main sensible conclusion in this context is that a lot of integration and interfacing work is in front of the people involved in Remote Handling business. It is certainly one of the few fields in which Systems Engineering will take even a bigger importance with the growing spectrum of the environment in which these techniques are required and the more and more complex mission to perform.

TELEMAN: A EUROPEAN COMMUNITIES ROBOTICS PROGRAMME FOR THE NUCLEAR INDUSTRY

Brian TOLLEY
TELEMAN Programme, DG XII
Commission of the European Communities
Rue de la Loi, n°200,B-1049, Brussels, Belgium

ABSTRACT

The TELEMAN Programme is a five year cost-shared research programme covering remote handling in hazardous and disordered nuclear environments. It is supported within the current research and development programme of the European Communities.

TELEMAN's strategic objective is to develop advanced teleoperators that respond to the needs of the nuclear industry. Its technical objective is to strengthen the scientific and engineering bases upon which the design of teleoperators for use throughout the nuclear industry rests. This will be done by providing new solutions to problems of manipulation, material transport and mobile surveillance in nuclear environments and by demonstrating their feasibility.

Motivation for such a programme lies in the potential teleoperators have to improve the separation of workers from radioactive equipment. This technology will also enable plant operators and public authorities to deal more effectively with nuclear abnormal incidents and increase gains in productivity, mainly in the repair and maintenance area.

Community support is justified by the cost of the reliability and autonomy required for the nuclear teleoperator, the need to rationalise European R&D competence in an area of increasing industrial potential and a common interest in coherent responses to emergencies.

1.- INTRODUCTION

Remote handling has always been an important part of nuclear technology characterised by the exercise of considerable mechanical engineering ingenuity under hostile conditions and a continuing compromise between simplicity and automation. Computer assisted teleoperation in which man shares the functions of control with a machine offers new options: the ability to work faster and longer than the human operator; relief from the

M. C. Becquet (ed.), Teleoperation: Numerical Simulation and Experimental Validation, 7–13.

drudgery of repetitive operations; a big increase in the consistent quality of work, particularly in unpleasant environments; machines able to maintain machines and the transfer of complexity from operator to a control system with which he can communicate in something approaching his own language.

The TELEMAN Programme has the following long-term and operational objectives :

-To encourage the development of an EC capacity to manufacture TELEMAN equip- ment for nuclear maintenance, inspection, repair, renewal and accident intervention
-To reduce the radiation exposure of workers
-To ensure effective research investment
-To see that researchers and other workers with the necessary skills are available
-To make sure that the needs of industrial users and manufacturers are taken into account and met as far as possible through integration of relevant technologies
-To build a network of collaborating research groups
-To raise awareness of the Technology
-To produce components and sub-system building blocks from the 1st series of projects
-To incorporate these building blocks into machines for user testing in realistic environments
-To encourage high quality participants to obtain only very high quality research projects

2.- NUCLEAR REQUIREMENTS OF TELEMAN

The range of radiation fields experienced in the nuclear industry makes it convenient to consider three classes of application :

-applications in which a reduction of doses to the nuclear workers is achieved by avoiding many small exposures to radiation
-applications where several men or teams of men are needed to avoid individual dose limits being exceeded
-applications in which man access is inconceivable

TELEMAN addresses the two latter groups of applications, those characterised by a demand for flexibility and by uncertainty within hazardous environments. Complexity, difficult communications, poor vision and the shortcomings of man as a controller of processes with many degrees of freedom have led to the concept of computer assisted teleoperation where the operator can call up a number of autonomous functional packages to execute specific jobs. The computer can, in turn, given suitable inputs from sensors or position encoders, help the operator to avoid collisions, illegal or dangerous actions.

3.- PROGRAMME STRUCTURE AND CONTENT

TELEMAN will address basic problems associated with remote operations using computer assisted teleoperators with increasing degrees of autonomy and invest most of its resources in the provision of robust generic building blocks for incorporation in a variety of specialised machines.

The research products will be incorporated into research machines that exist or are under development in its contractors' laboratories. Analysis of the behaviour of research machines will indicate how they will need to evolve to meet commercial demands and thus lay the groundwork for product development and demonstration programmes for a competitive environment. Evaluation of the products by users will include reliability assessments, mission simulations and involve consideration of testing strategies.

The Programme structure is based on the following areas of work :

3.1.- Component and Sub-system Development

Examination of the machine functions required for a wide range of teleoperator applications reveal the need for a relatively small number of basic technologies, i.e. those connected with perception of surroundings, intelligent job or route planning and decision making, power transmission, locomotion, manipulation, control and communication.

3.2.- Environmental Tolerance

In order to work in nuclear environments teleoperators will need to tolerate radiation, heat, corrosive chemicals, water, steam, dust, electromagnetic interference, vibration and decontamination. TELEMAN will provide design methods and data on environmental tolerance and failure modes.

3.3.- Research Machine Projects

Research machines are experimental machines that precede prototypes and demonstrate the integration of the disciplines of informatics, engineering, ergonomics and psychology involved in teleoperator construction into neat serviceable packages. They also serve as test-beds for novel components or control systems and essential aids for the training of engineers in teleoperation and robotics.

The Programme should be focussed on the following classes of teleoperator :

i)Improved light, mobile machines that can be used to collect information, relay signals, provide alternative viewpoints or serve as an extra hand

ii)More robust, mobile machines for maintenance, repair, post accident intervention or decommissioning including manipulators attached to vehicles, gantries and teleoperated cranes

iii)Improved long reach devices able to work through restricted apertures in a containment boundary

3.4.- Product Evaluation

This task has three objectives :

- -To involve end-users of teleoperators in TELEMAN by inviting them to test and evaluate concepts developed in the laboratory. This includes the setting up of relevant test strategies and designation of suitable test facilities
- -Monitoring of developments capable of influencing trends in teleoperator development
- -Monitoring the status of tools for the assessment of performance, reliability and user- friendliness

4.- IMPLEMENTATION

The Council of the European Communities officially adopted the Programme on 18 July 1989 with a total Community funding of 19 million ECU.

A TELEMAN Users Group has also been set up to ensure the interests and views of users are considered and is involved in various studies such as the definition of functional requirements of different types of machines.

A first call for research proposals covering the areas in items 3.1. and 3.2. above was published in September 1989. 41 proposals were received varying in size between overall costs of 180.000 ECU and 2.600.000 ECU, and between less than 2 man-years and almost 30 man-years of work. The total requested Commission funding was three times the available funds of about 9.000.000 ECU. There was a total of 157 participants in the proposals, making an average of about 4 participants per proposal. Following a selection procedure in which independent evaluators participated, 16 contracts (Tables 1 and 2) have been negociated with a total Commission funding of just over 8.000.000 ECU. Some of the contracts start as short duration feasibility studies which if successful will be allocated an extra sum of about 1.000.000 ECU.

It is interesting to note that 20% of the participants in the contracts are new to the Commission's DGXII existing shared-cost research activities. TELEMAN is unlike the majority of other CEC Research Programmes in that all of the contracts have been designed to be complementary in their contributions to the unifying goals of the Programme. Each contract will provide one or more "building blocks" to be proposed for incorporation into the various types of research machines under the 2nd phase of the Programme.

The relevant Second Call for research proposals covering the areas in items 3.3. and 3.4. above is planned for October 1991 with a Commission contribution of about 7.000.000 ECU.

TELEMAN is a new Community Research Programme of relatively modest size. However the response to its launch has been so positive as to ensure strong support for it and its possible development from European Community Member States.

TABLE 1: TECHNICAL COVERAGE

TECHNICAL AREAS	16 CONTRACTS															
	1	2	5	7	8	11	14	15	17	18	20	22	26	29	34	41
1. MANIPULATOR - HAND										●	●					●
2. MANIPULATOR - ARM			●								●					●
3. LOCOMOTION - GANTRY				●							●		●			●
4. LOCOMOTION - VEHICLE					●		●	●								●
5. LOCOMOTION - LEGGED						●										●
6. OBSERVATION - IMAGES	●	●	●	●					●			●	●	●	●	●
7. OBSERVATION - RANGE AND SHAPE	●	●		●				●	●					●		●
8. MAN-MACHINE INTERFACE	●		●	●				●	●		●		●			●
9. MAPPING - WORLD MODELLING				●				●	●				●	●		●
10. MISSION PLANNING - CONTROL				●	●	●		●	●				●			●
11. COMMUNICATIONS	●			●	●	●		●			●	●	●			●
12. FAULT TOLERANCE NUCLEAR ENVIRONMENT	●			●		●		●		●	●	●			●	●

TABLE 2 : TELEMAN CONTRACTS

A complete list of all the complementary contracts within the TELEMAN Programme is given below. All of these contracts are interdependent in that they are intended to contribute building blocks which could be used in subsequent research machines projects to be selected under the second TELEMAN call for proposals.

TM No.	TITLE	PARTICIPANT	COUNTRY
1	**TELEMAN 01** : Advanced Radiation Tolerant Information and Sensor System for Teleoperation (ARTIST)	• IBP Pietzsch • AEA Harwell • KHG • Le Conoscope • NCSR Demokritos • Lens Innovation • Univ. Telecom	• D • GB • D • F • GR • GB • F
2	**TELEMAN 02** : Laser Scanning Telemeter, for Circulation in Hazardous Environment or for Space Sensoring, from a 2D TV Perception with Adjustable Window for the 3D Scanned Network (LAST CHESS PAWN)	• Bertin • EDF • Etnoteam • Initec • Univ. Ioannina	• F • F • I • E • GR
5	**TELEMAN 05** : Advanced Control System for a Master-Slave Force Reflecting Servo-Manipulator	• ENEA • Prima • Univ. Oxford • JRC-JET	• I • I • GB • GB
7	**TELEMAN 07** : Vision System and Sensors - Building Block B of "Components for an Intelligent Robot and Crane Control, Usability and Intelligent Teleoperation (Circuit)" (VISYS)	• CEN/SCK • CRIF/WTCM • KfK Karlsruhe • Univ. Oxford • AEA Harwell	• B • B • D • GB • GB
8	**TELEMAN 08** : Development of Functional Blocks to be Incorporated in a Mobile Vehicle that could be used to Navigate Autonomously in a Complex Environment like a Nuclear Plant	• Ansaldo • Univ. Genoa • Kath. Univ. Leuven	• I • I • B
11	**TELEMAN 11** : Study of an Autonomous Legged Robot for Intervention in Nuclear Installations (SHERPA)	• Technicatome • CEA • EdF • Univ. Libre Brussels • Univ. Karlsruhe	• F • F • F • B • D
14	**TELEMAN 14** : Technologies for Advanced Locomotion Systems (TALOS)	• CEA • KfK Karlsruhe • Cybernetix	• F • D • F
15	**TELEMAN 15** : Assistant Navigation Block (ANB)	• FHG-IPA • Framatome • Equipos Nucleares • KHG • Univ. Libre Brussels	• D • F • E • D • B

TM No.	TITLE	PARTICIPANT	COUNTRY
17	**TELEMAN 17** : Universal Knowledge Based Imaging System (UKIS)	• TNO-TPD • Dornier • Tecnomare • Ikerlan • TNO-FEL	• NL • D • I • E • NL
18	**TELEMAN 18** : Development of a Sensor Based General Purpose Prototype Gripper	• Tech. Univ. Delft • Vermaat • Univ. Libre Brussels • Borssele • NOK-Beznau • JRC-Ispra	• NL • NL • B • NL • CH • I
20	**TELEMAN 20** : Manipulator and Gantry Intelligent Control - Building Block A of "Components for Intelligent Robot and Crane Control, Usability and Intelligent Teleoperation (Circuit)" (MAGIC)	• AEA Harwell • Interatom • KfK Karlsruhe • Univ. Newcastle • Univ. Hannover • Nuclear Electric • Lapp	• GB • D • D • GB • D • GB • DK
22	**TELEMAN 22** : Development of an Intelligent Radiation Vision System	• Ciemat • AEA Harwell • Equipos Nucleares • Amys	• E • GB • E • E
26	**TELEMAN 26** : Simulation, Planning and Programming Environment - Building Block C of "Components for Intelligent Robot and Crane Control, Usability and Intelligent Teleoperation (Circuit)" (SIMPLE)	• KfK Karlsruhe • Nuclear Electric • Interatom • AEA Harwell	• D • GB • D • GB
29	**TELEMAN 29** : A Remote Measuring System for the Mapping of Hazardous Environments (HAZMAP)	• Univ. College London • Nuclear Electric • Cadcentre • Kotowski	• GB • GB • GB • D
34	**TELEMAN 34** : Radiation Tolerant CCD Camera (RATOCA)	• Sofretec • Thomson • Ciemat • AEA Harwell	• F • F • E • GB
41	**TELEMAN 41** : Environmental Tolerance, Reliability and Safety - Building Block D of "Components for Intelligent Robot and Crane Control, Usability and Intelligent Teleoperation (Circuit)" (ENTOREL)	• Risoe • CEN/SCK • Interatom • AEA Harwell	• DK • B • D • GB

PRACTICAL EXPERIENCE USING TELEOPERATED TECHNOLOGY: TELE-OPERATED DEVICES USED IN AN ACCELERATOR COMPLEX

Roger HORNE
Mechanical Technologies Division, Engineering Support Group,
Teleoperation Section, CERN,
CH-121 Genève 23, Switzerland

TELEMAN supports the development of teleoperated devices which could be used in "disordered environments". In clear,this could mean a Euro-T.M.I. or at worst, a Euro-Tchernobyl, where routinely-employed remote handling equipment could not cope.
The CERN accelerators have no potential to create such ultimately hostile environments. However, certain local zones contain experimental equipment which is highly radioactive and often unreliable. This must be installed, maintained, modified, exchanged or demolished, often at short notice, and as quickly as possible, using special tele-robotic devices.
Some missions are completely unforeseen. The operating team depends entirely on obtaining quality sensor information for making their decisions.
The devices must have the widest possible capability in order to implement those decisions.
These requirements parallel many of the performance objectives of TELEMAN.

Repeated experience in the CERN complex of tunnels and halls showed a need for a flexible multi-purpose capability involving 3 overlapping fields of remote operation, namely:- (1) Observation, (2) Manipulation, (3) Lifting/ Transport.
Therefore we focus our continuing development effort into three coherent projects, namely:- (1) Private Eye, (2) MANTIS and (3) ACOL Service Vehicle.
These devices are operated independently or together depending on mission requirements.
The 3 projects have some common performance objectives:

Rapid response. Such devices should be immediately available, even after long periods of inactivity, and be simple to operate.

They should be easy to transport deploy and retrieve, without damage.

Full mobility and sufficient autonomy, with reliable two-way communications, eventually over kilometre distances, through varied and difficult access routes.

This combination of objectives is an eternal compromise.

M. C. Becquet (ed.), Teleoperation: Numerical Simulation and Experimental Validation, 15–20.

MISSION SCENARIO (1).

A client group asks to examine damaged equipment in a locally highly radioactive zone 300m along an accelerator tunnel.
The teleoperator team are given site plans, drawings, and photos, suspected to be out of date.
This is confirmed later on the TV monitors of Private Eye during a first cautious entry through the "access route" into the "work zone". The team builds a reliable data-base of on-line video or graphic images, on tape or laser-disc, for collision avoidance as the mission continues, and for the client group to identify and examine their equipment in detail, and decide what action to take.

OBSERVATION (Private Eye series).

As seen from the above, a clear first objective is to move and orient a tv camera/zoom/pan/tilt/ package literally anywhere in a working environment.
At CERN, straight counterbalanced booms suspended from a bridge crane were able to usefully "fly" cameras around, but they occupied or swept too large a volume. Articulating arms were more versatile and required less parking space, particularly when the arm links were made telescopic. Such linear movements were found to be easier to control but heavier and more expensive than articulating movements. The value of having one TV system observing the safe movement of another was learned.
Changing to floor-running systems in 1976, CERN acquired its first "STEM" furlable element mast which when suitably modified, and fitted on an orienting base, became the first of the Private Eye series. The latest, P.E.3, (Fig.1) has a compact, battery powered radio controlled base vehicle from which a TV camera can be "flown" out 5 metres in any direction by a specially made STEM mast.
A second camera on the vehicle surveys the movement of the STEM mast. TV images from PE 3 are video-recorded, and can also be digitised and stored on laser disc. Typically the evolution of cracks, corrosion, leaks and debris are important. Other sensor information such as radiation levels, distance, humidity, temperature, gas concentrations will be displayed on-screen and thus also stored as evolving reference data.

MISSION SCENARIO (Continued)

The client group must retrieve their equipment from the tunnel within the next 12 hours, without damage either to it or adjacent structures. Their equipment has water, compressed air, heavy power, and instrumentation connections.
There are doubts about the connector sizes, the overall weight, and theposition of the centre of gravity.
The team decides to send in MANTIS to first disconnect the equipment and ensure free access for lifting it out.

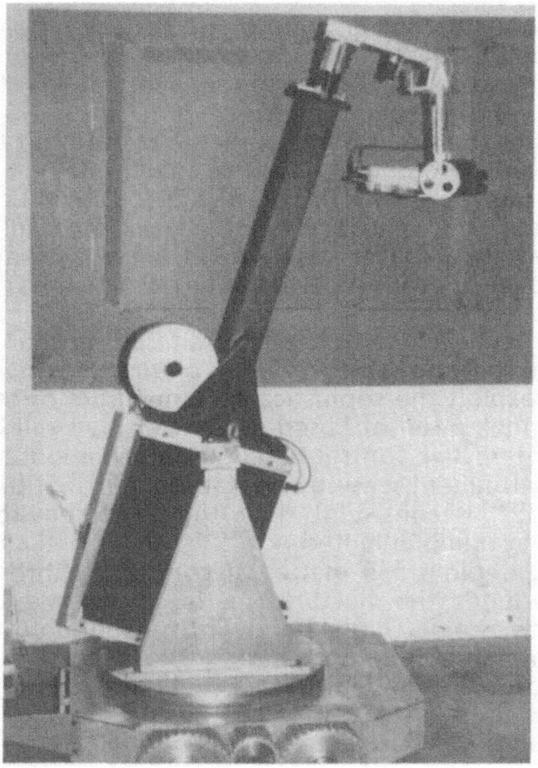

Fig. 1

Fig. 2

TELEMANIPULATION (MANTIS SERIES).

MANTIS was designed to "go anywhere and do anything, as a man could" inside CERN's complex of halls and tunnels.
Firstly, this infers maximum "compliance", the ability to adapt to the constraints of the working environment.
Secondly, it requires the best system available to transmit the dexterity and intellect of a human operator into the work zone.
When folded for transport MANTIS will pass through an access doorway 2m high and 1 m wide. The vehicle and articulating hydraulic crane serve to transport the servo-manipulator slave arms into a work zone, and "fly" them up to 10 metres over or around obstacles or debris to reach the work to be done.
This approach largely avoids the need for an "all terrain" or stair-climbing capability, and the consequent severe limits on size, weight and stability.
In order to fully exploit the sophisticated kinematics of the MANTIS crane safely inside a tunnel, a sensor based graphic display will be indispensable, eventually with interactive control functions which impose movement limits in response to environment sensing or database information.
An approach which is consistent with the CERN environment is under discussion, but not yet implemented.
MANTIS currently deploys 550 metres of combined fibre-optic and power cable which precludes the need for a large volume of batteries and guarantees totally stable high bandwidth communications which are essential for servo-manipulator operation.
The operational problems of deploying a cable are acceptable if only one of the teleoperators used on a combined mission is so equipped, and the deployment system is well designed.
The force-reflecting master-slave servo-manipulator, equipped with good quality video and audio, is a highly effective means of doing "man equivalent" work remotely.
Such a system allows a human operator in a control room to sense the distant work zone through his eyes, ears and hands, and to transmit precise forces and movements back into the zone, in real time, via the master and slave arms.
The word "transparence" is sometimes used to indicate the quality of a such a teleoperator system. Ideally the operator interacts naturally with objects or tools in the work zone. He is essentially unaware of the system though which he is working.
Digital techniques are now used to reduce "perceived friction" or "force reflection threshold" values to below 100 grammes with good stability.

In the scenario, given the variety of connectors and the time constraint, MANTIS, escorted and watched by Private Eye will enter the zone with a large supply of tools and try them successively until the right ones are found.
In principle, MANTIS could measure the connectors with a calliper gauge, so that the correct tools could be carried in by Private Eye, but in practice this process would take much longer, given the 300m access route and possible delays due to connector or tool damage. For power tools, MANTIS has its own controlled air and electric supplies, avoiding constraining links to local outlets.

Having separated the equipment connectors, and cleared away pipes, hoses and cables, and ensured that any fixing bolts are removed, MANTIS must move further along the tunnel to make room for the remote transport/lifting vehicle.

MISSION SCENARIO (FINALE).

The ACOL lifting/service vehicle, which has a hook load capacity at minimum extension of 7 tons, is driven remotely into the work zone escorted by Private Eye and carrying a special lifting attachment supplied by the client. When attempting to lift the clients equipment, said to weigh 3.5 tons, from its self aligning baseplate, the Private Eye camera shows the rear wheels of the vehicle lifting off the floor.

LIFTING AND TRANSPORT.

The applied forces and consequently the potential effects of an accident are greatest in this phase of the mission. The ACOL service vehicle resembles a compact 7 ton hydraulic crane on a forklift chassis.(Fig 2). It was conceived to install or exchange any of the beam-line elements in the Anti-proton collector target zone, but is foreseen to be used at longer ranges elsewhere in the complex when tunnel radio-communications achieve sufficient reliability.
The crane arm movements are simple slew and lift, with a telescopic jib and a rotating hook. The 1.4 metre wide times 2.5 metre long base vehicle has height of 1.9 metres and a turning radius equal to its length. It weighs about 2 tons.
With these kinematics, overall tunnel compliance is good, but with the important disadvantage that lifting substantial loads outside the vehicle's wheelbase/track area can lead to a capsize. This is discussed later.
The device is fitted with its own Private Eye (Mk1), able to "fly" a TV camera/zoom anywhere within 3 metres of the vehicle. A variety of attachments can also be fitted in place of the rotating hook, including a 5 d.o.f. robot arm, and a special hook for lifting concrete shielding blocks. All of these can be monitored by the P.E. camera.
The situation in the scenario is based on a case history and, in the "disordered" context should not be regarded as abnormal. After a thorough search with Private Eye to check that all fixing bolts had been removed, another attempt was made. This time, with the P.E. camera at floor level, it was detected that the equipment was not lifting vertically, and was jamming against its over-toleranced alignment pins.
It was found that some redundant components had been removed from the equipment after installation. Consequently, although the overall weight was reduced, the special lifting attachment was no longer in equilibrium.
The solution finally adopted was to reach across with MANTIS and "lean" on the end of the equipment to hold it in equilibrium until it lifted clear of its baseplate pins and was hanging free. The service vehicle could then orient the suspended load, using the rotating hook, so that the profile of the vehicle and its load was a minimum, reducing the risk of collision during the return journey along the accessroute, monitored by Private Eye.

As noted above, the question of safe operation under conditions of reduced vision or "awareness" is crucial. Following an incident in 1990, the service vehicle now has a simple PC based graphic display showing the kinematics of the vehicle in plan and profile and horizontal lines representing tunnel entrance and roof height at 2m and 3m. The position of the centre of gravity of the combined vehicle and load is indicated by a cross which must be kept within the depicted plan area to avoid capsize. This proved to be very effective in later operations, and work continues on applying similar techniques to MANTIS and Private Eye.

COMMENTARY

1) When shut down for maintenance, the induced radioactivity around 90% of the CERN accelerator complex is at a low level which allows controlled personnel intervention. In this proudly hi-tech and time-constrained physics environment, attitudes prevail to the effect that tele-operator development is an unnecessary nuisance. The result is that in order to be accepted, the teleoperators must demonstrate a capability similar to that of men, at low cost, and with the minimum of inconvenience to the physics programme.

2) Being essentially in competition with personnel intervention has pushed us toward some of the performance objectives of TELEMAN, under operational conditions. Such experience has much value, even without having ultimate resistance to radiation or all terrain vehicle capability.

3) New uses for some sort of rapid deployment ultra-versatile remote capability appear every week, particularly since the Gulf war. The compliance needs for different environments may change, but most of the technology remains the same. It is to be hoped that TELEMAN or its successor will continue to support the European effort to meet these demands.

ARTIFICIAL REALITIES TECHNIQUES FOR TELEOPERATION OF ROBOTIC SYSTEMS

DANIELE MARINI
Dipartimento di Scienze dell'Informazione
Laboratorio di Eidomatica
Università degli Studi di Milano
Via Moretto da Brescia, 9 - 20133 MILANO, ITALY
tel. +392 7575 267 - fax +392 7611 0556

ABSTRACT.
Recent advances in visual simulation and Man Machine Interface methods and tools have given rise to a new viewpoint in Human Computer Interaction: Artificial Reality. This viewpoint can be considered as a technical and scientific paradigm, well suited for the implementation of advanced teleoperation systems. In the paper a review of A.R. techniques is presented, with a particular emphasis to aspect relevant for teleoperation.

1. Introduction

Robotic Teleoperation is a very complex activity that requires a high degree of Human Computer Interaction, both in the planning phase and during operation for monitoring and control purposes. The presentation to the human operator of the environment and robotic system status must be maintained at a very high level of quality, in order to minimize human effort to interpret messages received by the computer system. The operator moreover must be supported by intelligent interfaces in order to send high level commands and messages necessary to plan and control the given system.

During robot task and mission planning it is particularly useful the possibility of simulating robot operation in order to verify correctness, robustness and completeness of the devised action plan.

During robot operation in field it is extremely important a continuous monitoring of robot action by the human operator in order to suspend automatic execution in case of risk or dangerous events, and recover the overall system.

All these requirements can be fulfilled developing software and hardware systems capable of high level interaction, graphical presentation and intelligent behaviour.

High level graphics interaction is based on the so called *telepresentation* and *telepresence*, meaning the tentative to mimic at the highest possible level external reality through specialized devices and software, aiming to reproduce at the operator site the same stimulus that he would receive in the operative area.

At such purpose it is necessary to model a robotic system and the operative environment at a very detailed level, considering geometric and morphological aspects, kinematic and dynamic properties, and even visual and other physical properties. A complete model can therefore be the basis of a complete simulation of the behaviour of the robotic system,

M. C. Becquet (ed.), Teleoperation: Numerical Simulation and Experimental Validation, 21–25.

while particularly suited devices can convoy to the operator a representation of the status and evolution of such modeled world.

Above described process is typical of the Artificial Reality systems, a relative new computer application area, whose aim is to bring a computer user in a non existent or remote world, allowing him to interact with it in the same way as such world would effectively exist.

In this paper we present a quick overview of fundamental concepts, devices and methods for A.R. systems, and we shall emphasize the aspects of major importance in the area of robotic teleoperation, and we shall see how A.R. methods and techniques can be useful for planning and controlling complex robotic devices.

2. Building Artificial World for Robot Teleoperation

An Artificial Reality World is composed of several parts, and A.R. simulation requires the solution of many different and complex programming and technological problems.

First of all it is necessary to build a geometrical representation of the Artificial World and of all the elements that will live inside it. This, at first sight, seems very close to CAD/CAM problems, that still require the construction of geometric representations. There are nonetheless many differences, due mainly to the different purposes of geometric modelling in the two contexts: a geometric model for design purposes will require particular details for verifying its shape, for numerical computation of mass and volume properties, and for developing its construction schema. On the contrary a geometric model for an A.R. application stresses its functional and operational characteristics, e.g. its structuring into independent but correlated parts in the case of a moving and articulated equipment. Moreover in A.R. context the computational problem is mainly a problem of simulation, therefore the geometry data base must be organized in order to minimize time spend in traversing the data structure and computing most relevant numerical tasks, e.g. collision detection, or kinematics and dynamics simulations.

The geometric model of a robotic system is composed of different interconnected part, whose relative movements are binded in a kinematic chain; a hierarchical representation is the best solution for describing such kind of models, and allows to control in detail the propagation of geometric transformation generated by a given movement sequence.

The geometric representation of the A.R. world and components is the ground step in building a simulation environment. As a second step the simulation must be planned and simulation tools must be devised. In this context we consider simulation as a set of utility programs that allow the A.R. world to evolve obeying physical laws with a behaviour as demanded by a human operator or by a previously defined script. We can understand better what are the conceptual and practical problems involved considering the A.R. world as a theatre or movie stage, where actors will play their role following a general script, detailed for each actor, and where a director will manage the whole play. Staging, from this viewpoint, is the same activity as modelling the A.R. world, but modelling is not complete unless one describes actor scripts, and how actors will interact each other. The final result will be playing the script, which is the simulation.

Before we analyse with greater detail what are the issues relative to "directing" the play and displaying the simulation (i.e. operator interaction with the A.R. world and A.R. visualization etc.), let's spend some more words on staging the A.R. world and preparing models in order to allow them to interact following their scripts.

The metaphor of "scripts and actors" appeared in computer science since at the end of 70's, when Carl Hewitt first tried to describe in this way communication and synchronization problems of concurrent processes. From computer science viewpoint the interaction between actors is a problem of exchanging messages and controlling that all events will come in their correct timing sequence. Present day programming technology allows to describe this kind of concurrent processes at a very high level of abstraction, recurring to the so called Object Oriented Programming, a programming methodology that allows to encapsulate actor behaviour into "objects" with their behaviour (methods), communicating each other through messages; when messages are received by an object, its reaction obeys the particular method associated to the received message. Following O_O methodology, actors can be described and programmed with relative easiness.

In order that each element of the A.R. world plays its correct role it is necessary to model it also from its physical viewpoint and it is necessary to have computer software solving in an efficient way direct and inverse kinematic and dynamics of rigid bodies. In such a case it will be possible to assign to each element its task in terms of, for instance, a trajectory to be tracked in a given time span with some given constraint, and leave to the system to develop time equations and computation of intermediate positions, checking presence of obstacles. An intelligent behaviour can be devised, allowing the computer system to cope with obstacles and other anomalies. At this purpose techniques from artificial intelligence can be adopted, increasing therefore the abstraction level where the human operator can interact with the system during planning or monitoring the action of the robotic system.

3. Visual Simulation for Robotic Teleoperation

A critical role is played by visual simulation in A.R. simulation. Indeed vision is the principal perception system of human beings, therefore image generated by computer systems are the most important way of presenting results of a numerical simulation. In the context of robotic teleoperation, visual simulation can be limited to the photorealism level, while particularly relevant is the capability of presenting images generated at a real time rate.

The degree of realism of a computer generated image can vary from lowest level of perspective line drawing of a geometric scene, up to photorealistic picture, almost indistinguishable from real photographs or TV displays. A choice between these two extreme levels in realistic rendering of geometric scene must be done, coping with two opposing requirements: from one side the readibility of the generated image, that must be unambiguous and immediately understandable; the speed of rendering, on the other side, that must stay within 25-30 images per second, in order to reach smooth rendering of animated objects.

The problem becomes more complex considering that the animation of a geometric scene, in robotic teleoperation, must not be too approximate or qualitative; on the contrary must be the result of a numerical solution to kinematics and dynamic equations, describing the behaviour of the robotic system in an accurate modelled physical world. More and more complex becomes the problem if the task of a robotic system is computed in real time, during the simulation, and AI techniques are solved concurrently with physical properties computation and image rendering.

State of art allows to render in real time rather complex scene with simple level of realism, the so called smooth shading. In order to compute in real time both images and physical properties it is moreover necessary to adopt advanced computing architecture, based on parallel computing environment, devoting some processors to numerical computation and others to image generation. A parallel approach is particularly interesting if stereopresentation are required. In this case two different views of the simulated scene must be computed for the two observer's eyes, and send to two different monitors or on a special stereomonitor. This approach allows to increase the quality of visual simulation, allowing the observer to perceive with greater easyness depth and relative position of the objects of a scene.

Above discussion is particularly suited for the planning phase of a robotic systems; rather different are the problems posed during monitoring and teleoperating a robotic system. In this second case the robot is working under control of its proper computer system, while the human operator has the responsibility of verifying the correctness and adequacy of the executed task with the formerly planned one, and get direct control in case of anomalous situations, critical stages or emergencies.

In such a case the role of vision is completely different: the human operator must be continuously prompted with picture taken on the operating site from, e.g., special TV cameras, presenting him either the view from the robot or a general view of the environment.

These devices allows the operator to "feel" like he were personally present in the operative site, realizing the so called telepresence and telepresentation of the operation site.

4. Human Computer Interaction in Teleoperation

The fundamental problem of human computer interaction has two principal facets. First of all it is not only necessary to present to the human operator a statisfying visualization, it is also necessary to expand the interaction through other sensorial apparatus. Therefore special devices are being studied and prototypes are available in order to facilitate kinestesic and tactical stimulus to be emulated, as well as audio and speech synthesis and recognition. Also visual stimuli can be largely enriched using stereovision, that facilitates the perception of depth and dimensions. The aim of this approach is to envelop the human operator in a totally stimulating environment, where he can no more distinguish between real and fictitious facts.

The second facet of HCI concerns the logical and conceptual aspect of interaction. The availability of above quoted devices allow to design new languages to support human computer dialogue. It is no more necessary to type in commands, even classical menus can be passed, adopting the well-known paradigm WYSYWYG (what you see is what you get), adopting iconic representation of commands, visualizing each component of the artificial world with highly realistic depiction. Therefore it is possible to the user to interact with the displayed object sending it a particular message and see immediately its reaction.Using data glove devices and speech driven command interpreters an operator can select an object and give it a command. As an example this kind of interaction technique allows to develop in a very easy way a new plan for a robotic system. Indeed the operator can connect his own hands with the end effector of a robotic arm, and pull it around in the virtual space instructing the system to a new task. What will be going on inside the A.R. system will be an evolution of the geometry of the robotic arm, which will obey both kinematic and dynamic bindings, via a real time solution of inverse dynamics, and at the same time, a visual simulation process will display to the operator, computer generated images with the desired level of realism, the movement of the system inside its operative environment.

A hierarchical decomposition of the levels of operation of a robotic system can moreover improve operators capability of planning its missions. Developed in a context of space research a proposal NASREM suggests to subdivide the architecture of a robotic system into three principal areas: sensory, world modelling, task decomposition. The sensory are concerns all sensory equipment that allow the robotic system to build a representation of the external world; the world modelling consists of a representation of the world while task decomposition describes the task to be performed by the robot. All the three areas are moreover decomposed into six levels of details: the lowest level is close to hardware devices, e.g. in the task decomposition is the electrical impulses that drive actuators; the levels increases in abstraction up to highest, where sensory processing, world modelling and task decomposition are described at a human operator language. This decomposition allow to build plans for new operation of a robotic system starting from elementary bricks that are assembled constructing very complex planes.

The approach of decomposing in hierarchical levels of abstraction the model of a robotic system and of its interaction with an operator, allows to simplify the command and control of the robotic system, in particular during the stage of task planning: at lower abstraction levels the operator will interact with A.R. devices controlling directly (virtual) robot action, while at higher abstraction levels the operator will interact with high level commands given with speech or other devices.

5. References

Badler, N.I. et alt. Ed's *Making them Move, mechanics, control, and animation of articulated figures*, Morgan Kauffmann, New York, 1991

Cox B., *Object-Oriented Programming,* Addison Wesley, 1986

Davis J., *New Interaction Media: Voice Input and Ouput*, Tutorial Notes n. 24, CHI 90 Empowering People Conference, Seattle, 1990

Deyo R., Ingrebtsen D., *Notes on Real Time Vehicle Simulation*, Tutorial Notes n.29, SIGGRAPH 89 Conference, Boston, 1989

Ferrario M., Marini D., *Computer e Immagini*, Ghedini Editore, Milano, 1988

Foley J.D. *Interfaces for Advanced Computing*, Scientific Americn, 257(4), 126-135, 1987

Goldstein H., *Classical Mechanics,* Addison Wesley, Reading (MA), 1980

Krueger, M. *Artificial Reality*, Addison Wesley, NewYork, 1991

Magnenat-Thalmann N., Thalmann D., *Principles of Computer Animation*, Springer Verlag, Tokio, 1985

Marini D. Ceriani S. - *Architetture distribuite per la Sintesi di Immagini col metodo di Ray Tracing* - PIXEL anno 9 n.9, 1988

Marini D., *Il fotorealismo per la restituzione di oggetti e scene*, Atti del Workshop Conoscenza per Immagini, CNR, Roma, Novembre 1989

Okoshi, T. *Three Dimensional Imaging Techniques*, New York, Academic Press, 1976

Peter B. Denes and Elliot N. Pinson, The Speech Chain: The Physics and Biology of spoken language, Anchor, 1973

Reid L., Nahon M., *Flight Simulation Motion-Base Drive Algorithms, Part 1 and 2,* Technical Reports 296 e 307, University of Toronto, Aerospace Studies, 1986

Schank R., *The Cognitive Computer,* New York, 1988

Uttal W., *I Teleoperatori*, Le Scienze, 285, 68-73, 1990

Watt A., *Fundamentals of Three Dimensional Computer Graphics*, New York, 1989

Wenzel, E.M.and Wightman, F.L. *A virtual acoustic display for conveying three-dimensional Information*, Proceedings of the Human Factors Society 32nd Annual Meeting, Anaheim, 1988

ROBOT MOTION PLANNING: A SURVEY

Carme TORRAS
Institut de Cibernètica (CSIC-UPC)
Diagonal 647
08028-Barcelona
SPAIN
e-mails: torras@ic.upc.es, eacarme@ebrupc51.bitnet

ABSTRACT. The approaches to robot motion planning can be classified into two categories: algebraic and heuristic. The former use *computational geometry techniques*, while the latter rely on *heuristic search*. In this paper, the most representative works within these two categories are reviewed, their main advantages and shortcomings being highlighted.

1. Introduction

Robot motion planning refers to the process of determining a sequence of robot movements (a plan), whose execution will lead to the attainment of a given robot configuration (the goal). The essential trait of planning is that it takes place before the actual execution of the movements, thus entailing the capacity to generate and manipulate internal representations of both the movements and the environment.

The environment and the robot movements need not be represented separately and, in fact, one of the key notions in motion planning is that of *configuration space*, which ties up both representations. Although this notion as such is that of a "state space" and thus it is quite old, the term itself was firstly introduced by Lozano-Pérez and Wesley (1979) and subsequently developed by Lozano-Pérez (1983). Most often it is taken to be the *space of degrees of freedom of motion* in which a given motion planning problem is to be solved. However, when the robot has non-holonomic joints (Laumond 1987; Laumond et al. 1988), the dimension of configuration space is larger than the number of degrees of freedom, and some infinitesimal variations of the configuration parameters cannot be achieved. Only holonomic systems will be dealt with in this paper.

Depending on the internal representations used and the algorithms employed to manipulate them, motion planning approaches can be classified into several categories and subcategories. *Algebraic approaches* represent configuration space explicitly by using continuous functions and use computational geometry techniques to plan motions. *Heuristic approaches*, on the contrary, usually rely on a discretisation of configuration space, which often is only implicitly represented, and use heuristic search to plan motions.

M. C. Becquet (ed.), Teleoperation: Numerical Simulation and Experimental Validation, 27–39.

These two categories are further divided in the paper. Thus, in Section 2, the main algebraic algorithms for robot motion planning are described and categorised as falling within the retraction approach or the decomposition approach. Next, in Section 3, the main heuristic procedures for robot motion planning are also reviewed and classified according to their use of either local or global heuristics. In each of these sections, the main advantages and shortcomings of the different approaches are discussed. Finally, in Section 4, the main conclusions emerging from this survey are made explicit.

Note that approaches to *task planning*, whose goal is to schedule operations so as to accomplish a given task, are excluded from this survey and the interested reader is referred to (Torras 1989).

2. Algebraic Approaches

There are two basic ways of modelling configuration space algebraically (Whitesides 1985; Yap 1987): one is to represent the "free-ways" between obstacles (by using, for instance, the Voronoi diagram) and the other is to represent the boundaries of the obstacles (through the use of some cellular decomposition). The motion planning algorithms developed using the former type of modelling fall within the *retraction* approach (Ó'Dúnlaing et al. 1983), since they "retract" free-space to only the middle line of the free-ways and then plan motions only along this line. Those that use the latter type of modelling follow a *decomposition* approach (Schwartz and Sharir 1983), and they plan paths through the graph of cells in the decomposition.

2.1. RETRACTION APPROACH

The key data structure in the retraction approach is the Voronoi diagram. This diagram is the locus of points that are equidistant from at least two obstacles. As stated, this definition applies to both physical space and configuration space. In the latter case, it is usually referred to as "generalized Voronoi diagram". For the particular case of a planar environment with polygonal obstacles, the Voronoi diagram is a planar network of straight and parabolic arcs (Figure 1a), which can be constructed in time $O(n\log n)$, where n is the number of vertices in the environment. Note that, in the case of a disc moving through this environment, configuration space is also planar with polygonal obstacles and thus the construction of the Voronoi diagram has the same complexity. For more complex moving bodies (e.g. a ladder) or more complex environments (e.g. 3D), the dimension of the Voronoi diagram increases and the complexity of constructing it increases as well (Ó'Dúnlaing et al. 1987).

The idea underlying the retraction approach is that there exists a continuous obstacle-avoiding motion between two specified configurations if and only if there exists another obstacle-avoiding motion between the same two configurations that, except for its initial and final phases (in which the robot moves to and from the Voronoi diagram), lies entirely on the Voronoi diagram. Note that the paths generated using this approach are the safest ones, since they evolve as far as possible from obstacles.

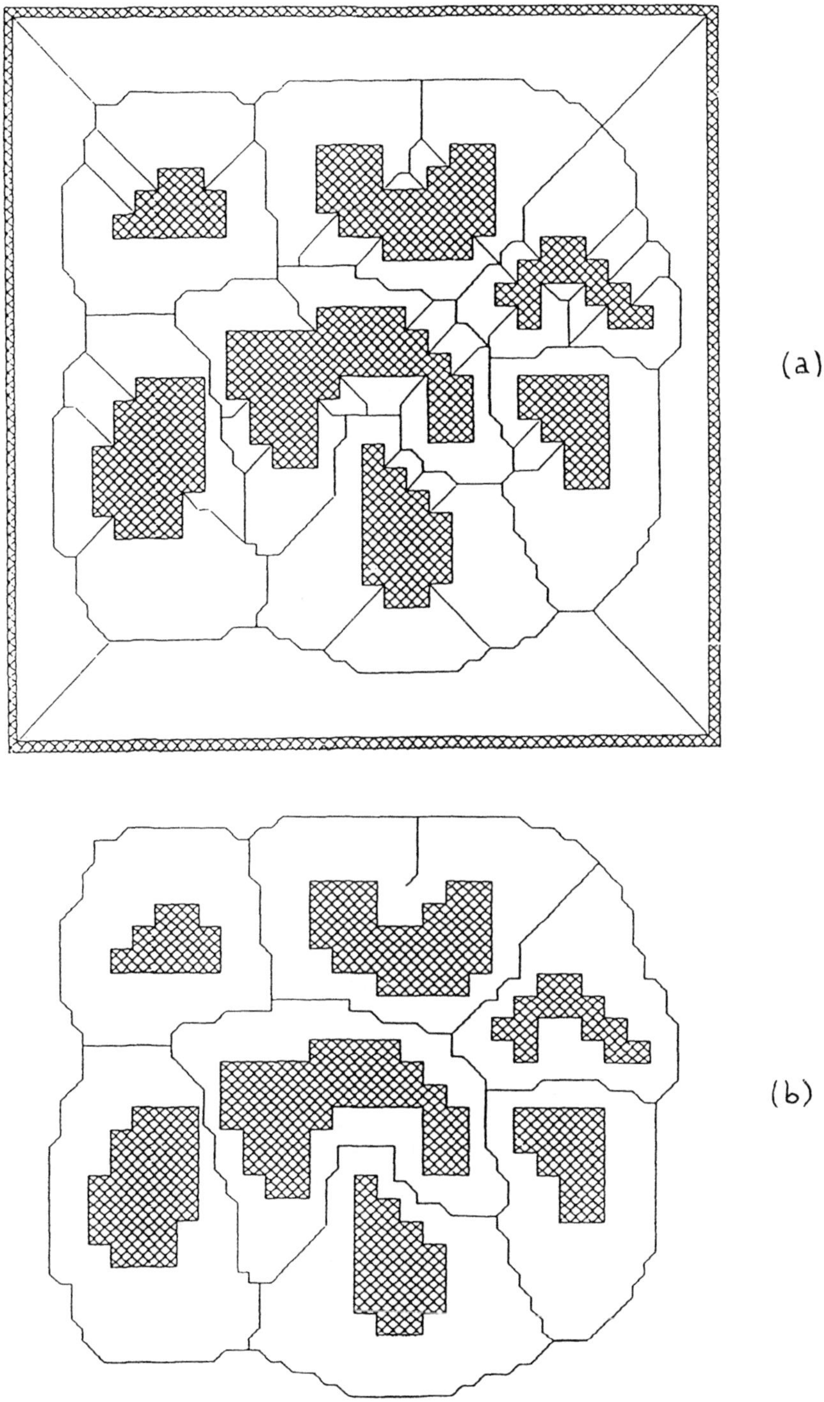

Figure 1. (a) Voronoi diagram of free-space. (b) R-MAT model of free-space.

Usually the construction of the diagram determines the complexity of the path planning algorithm, since it constitutes the most time-consuming subprocess. There have been attempts to simplify the Voronoi diagram in a way that retains as much clearance and connectivity information useful for path planning as possible. One such attempt has been in the direction of lowering the algebraic complexity of the surfaces defining a Voronoi-type free-space model (Canny and Donald 1987), while another consists of eliminating some irrelevant branches (e.g. those emerging from corners) from the representation (Ilari 1987) (see Figure 1b).

2.2. DECOMPOSITION APPROACH

The simplest algorithm following the decomposition approach is that developed by Lozano-Pérez and Wesley (1979). They consider the case of translational motion of a polygonal robot amidst polygonal obstacles. In this case, the obstacles in configuration space (called C-obstacles) are also polygons obtained by growing the obstacles in physical space according to the shape of the robot (Figure 2a), so that each obstacle-avoiding translational motion of the robot in physical space corresponds to the trajectory of a point in configuration space. Free-space is modelled in this work by means of a *visibility graph* (Figure 2b), whose nodes are the vertices of the C-obstacles and whose arcs link pairs of vertices that "see each other" (i.e. the straightline from one to the other does not intersect any obstacle). Path planning is then reduced to finding paths in this graph.

The construction of C-obstacles has been addressed in several other particular cases, namely for translational planar motion when both the robot and the obstacles have algebraic curve boundaries (Bajaj and Kim 1987), for both translational and rotational motion of a polygon amidst polygons (Avnaim and Boissonnat 1988), for 3D translational motion of a convex object amidst convex obstacles with algebraic surface boundaries (Bajaj and Kim 1986), and for the motion of a sphere among nonconvex obstacles with algebraic surface boundaries (Bajaj and Kim 1988). In all these cases, the configuration space is of dimension at most 3 and the C-obstacles are defined by algebraic surface patches together with a graph describing their adjacencies.

The C-obstacles arising in the case of a rigid robot navigating in 3D space are subsets of the 6-dimensional configuration space $\Re^3 \times SO(3)$, and their boundaries consist of patches of hyper-surfaces of dimensions less than 6. The general 6D motion planning problem has been addressed by Schwartz and Sharir, relying on the cylindrical algebraic decomposition of free-space (Schwartz and Sharir 1983). This is a decomposition into cells, each of which is characterized by a formula in predicate calculus, i.e. logic ANDs and ORs of predicates, whose variables are the degrees of freedom of motion (dimensions of configuration space). Schwartz and Sharir proposed an algorithm which is polynomial in the number of constraints that characterize the environment and doubly exponential in the dimension of the configuration space (Schwartz and Sharir 1983). The aforementioned decomposition is an extremely fine stratification and Canny, using the coarsest possible stratification of free-space, has been able to lower the complexity of general motion

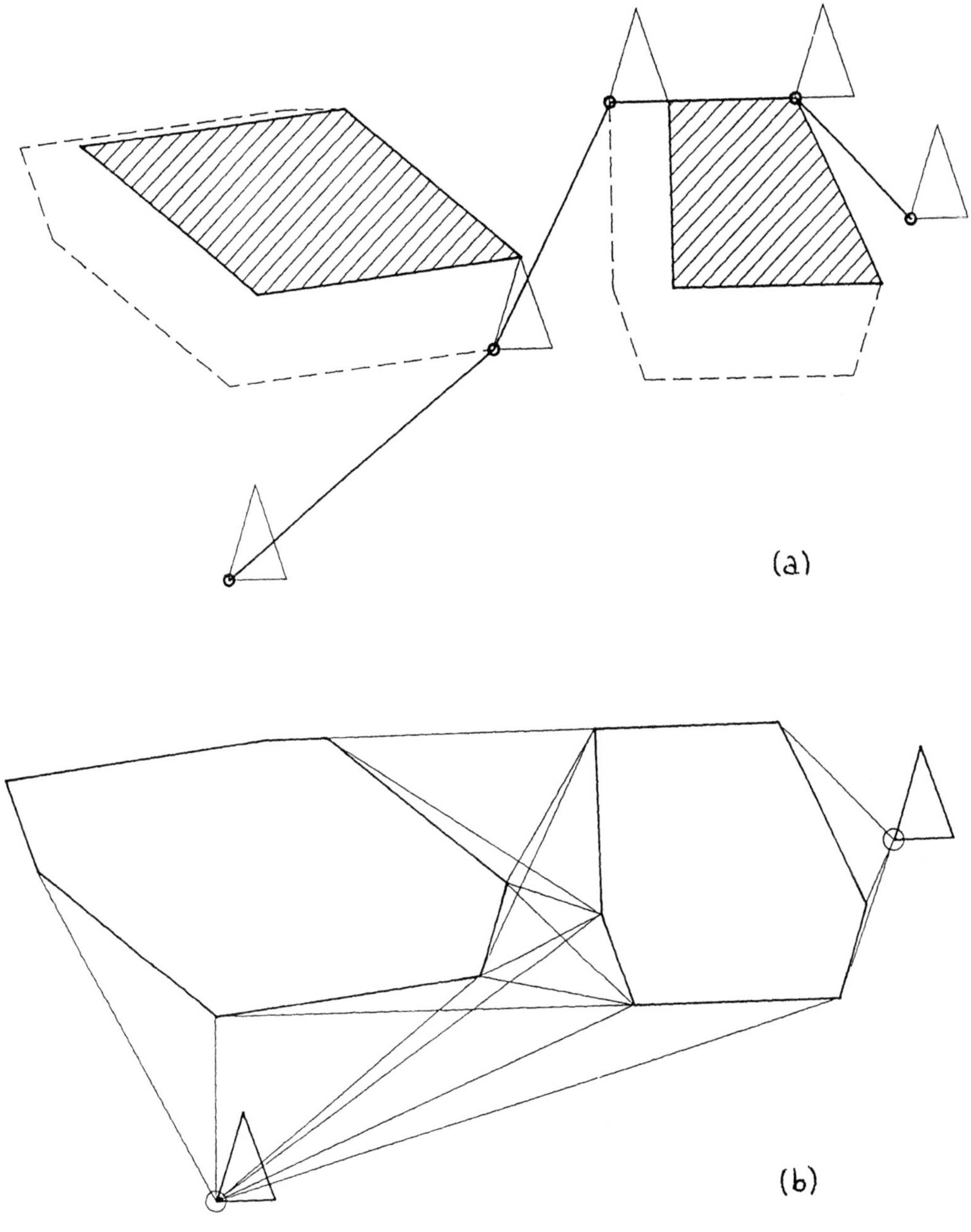

Figure 2. (a) C-obstacles obtained by appropriately growing obstacles in physical space. A continuous line that avoids C-obstacles corresponds to an obstacle-avoiding motion for the mobile body. (b) Visibility graph in C-space for the environment in (a).

planning to single exponential in the dimension of configuration space (Canny 1988). This is the best that can be done in the general case, since this last algorithm is worst-case optimal.

2.3. DISCUSSION

The main drawback of algebraic approaches is the need to effectively construct either type of model of configuration space (Voronoi diagram or cellular decomposition). This is a computationally intensive process, which constitutes the bottleneck for these approaches to be of practical use. In addition, algebraic approaches are conceptually complex, which makes their programming difficult, and of course are only adequate for off-line processing.

Only in the very restricted circumstances where the complexity of both the robot and the environment is very low may the problem be tractable using these approaches. For a disc moving amidst polygonal obstacles, there exists an algorithm based on the Voronoi diagram that takes $O(n\log n)$ time, n standing for the number of edges of the obstacles (Ó'Dúnlaing and Yap 1985). On the other hand, for a polygonal robot moving amidst polygonal obstacles, there exists an algorithm based on a cell decomposition of configuration space (Avnaim et al. 1988) with a tractable complexity of $O(n^3m^3\log nm)$, n and m standing for the number of edges of the robot and the obstacles, respectively.

3. Heuristic Approaches

Heuristic approaches rely on search processes, which are carried out either in physical space or in configuration space. In this section, they are classified according to the type of the heuristics used, either local or global.

3.1. LOCAL HEURISTICS

3.1.1. *Search in Physical Space.* The local heuristics used in this context are usually problem-specific (Udupa 1977; Luh and Campbell 1981), i.e. they depend heavily on the robot considered and the type of environment.

A differenciated approach within this category is that of carrying out gradient-descent on a potential field (Khatib 1986). Here the heuristic is the potential field resulting from associating an attraction force to the goal and repulsion forces to the obstacles. The main drawback of this approach is that the robot may get trapped into local minima of the potential field. To overcome this drawback, a Monte-Carlo technique that escapes from local minima by executing Brownian motions has recently been proposed (Barraquand and Latombe 1991).

3.1.2. *Search in Configuration Space.* Contrarily to the main trend in the preceding subsection, local heuristics used to guide search in configuration space can be of a general nature, since the shape of the robot is not considered separately now, but it is embedded in the shape of the C-obstacles.

A search process in configuration space guided by local heuristics of the general nature above is proposed in (Donald 1984, 1987). This is the heuristic path planning approach closer to algebraic approaches, since the search takes place on an algebraic model of configuration space of the cell decomposition type, and the heuristics are implemented through local experts that evaluate the shape of the boundary of a C-obstacle (e.g., intersection of two or more surfaces, degree of the surface) in a neighbourhood of the point under consideration, in order to propose the direction of search most likely to lead to a solution.

Up to our knowledge, this is the only heuristic approach to path planning that is *complete*, in the sense that it is guaranteed to find a solution when such a solution exists. Most other approaches rely on approximate models (e.g., discretised models) of either physical space or configuration space. This obviously introduces a degree of error: some feasible solutions may be lost and, on the contrary, some unfeasible solutions may be accepted as correct up to the resolution of the representation.

The search processes based on local heuristics, as those described above, lack knowledge of the connectivity of free-space, and thus cannot avoid getting trapped in pursuing unuseful explorations. To palliate this shortcoming, some algorithms have been proposed that use global heuristics based on a model of the connectivity of free-space.

3.2. GLOBAL HEURISTICS

3.2.1. *Search in Physical Space.* One algorithm that uses global heuristics in this context relies on a characterization of the "free-ways" between obstacles by means of generalized cones (Brooks 1983a). Thus, it follows a similar line of thought to that of the retraction approach, but confined to physical space. Each point in the middle line of a free-way has associated a set of permitted orientations for the robot. Path planning then consists in linking middle lines of free-ways, in such a way that the robot orientation in the intersection point between two such middle lines is permitted in both free-ways. This approach has been extended to deal with articulated arms (Brooks 1983b). A related algorithm closer to the decomposition approach relies on a representation of physical space as an adjacency graph of convex polygons (Chatila 1982).

3.2.2. *Search in Configuration Space.* One algorithm (Gouzènes 1983) that consists of a search in configuration space guided by global heuristics relies on a tree of collision-free zones (Figure 3). Each zone (tree leaf) is defined by as many coordinates as levels down the tree it lies on, taking the root to be level zero. This free-space model is built iteratively and ends up constituting a hierarchy of collision-free zones. The lower is a zone in the hierarchy, the more the number of coordinate checkings required to determine that a configuration in this zone is collision-free. Similar hierarchical decompositions of free-space are those based on octrees (Faverjon 1984) and on a recursive division into rectangloids (Brooks and Lozano-Pérez 1985).

The approach proposed in (Ilari 1987; Ilari and Torras 1987, 1990) slightly differs from those above in that the search space and the space whose connectivity is exploited

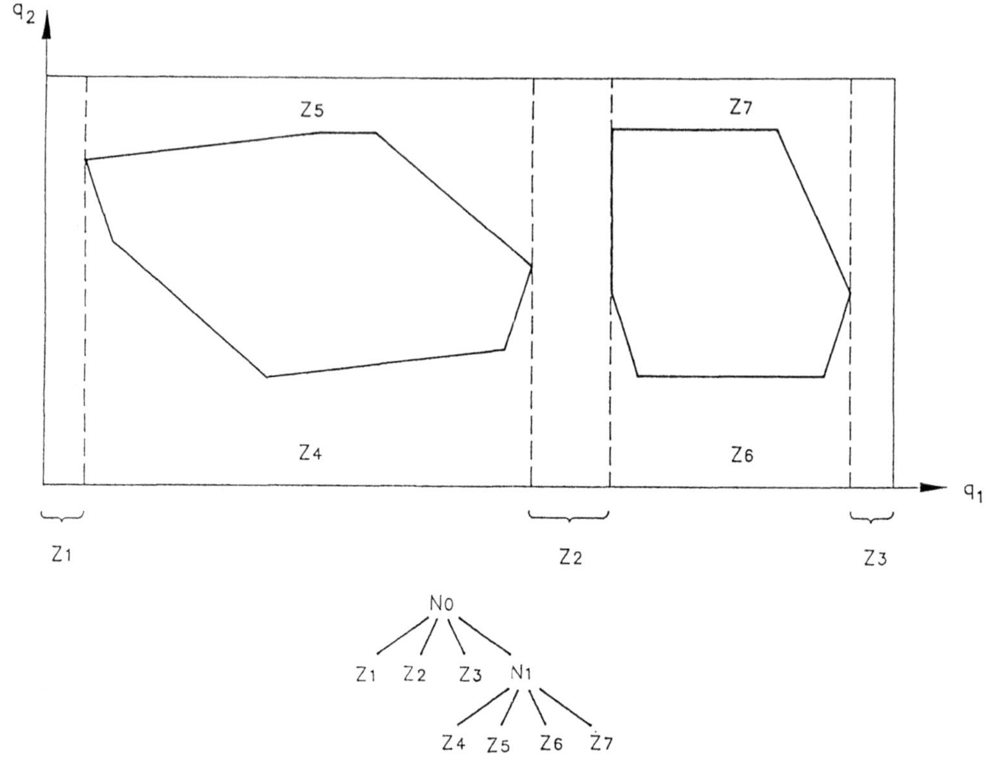

Figure 3. Construction of a tree of collision-free zones.

by global heuristics are not the same. The former is configuration space, while the latter is physical space. First, a model of physical free-space, based on the Voronoi diagram, is constructed and used to find a global path (Figure 4a). Next, a search process in configuration space —represented as a discrete n-dimensional grid of cells— is carried out (Figure 4b). This process uses two kinds of heuristics based on the global path: one that guides the evolution of the cartesian degrees of freedom of the robot, and two more that guide the evolution of the rotational degree of freedom. These heuristics lead to a speed-up of the order of 10^2 in relation to other approaches using the A^* algorithm.

3.3. DISCUSSION

In general, search algorithms using global heuristics have a big initial computational cost derived from the process of building the global model. However, once the model has been built, obtaining the sought solution path has a low computational cost which slightly augments with the complexity of the problem tackled. Search algorithms using local heuristics, on the contrary, usually require a low initial computational effort. However, the cost of converging to a solution (the search effort) rises quickly with the complexity

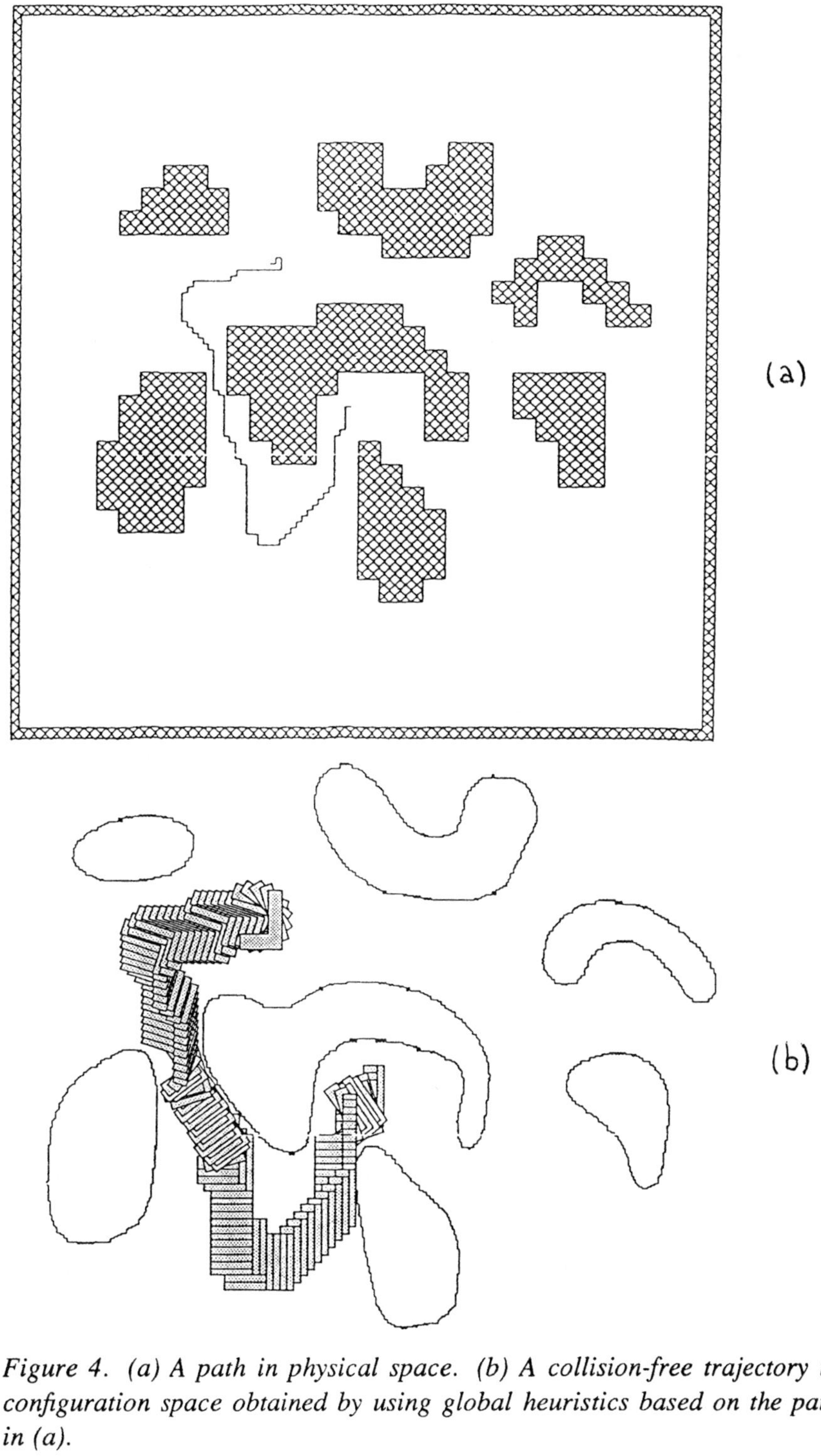

Figure 4. (a) A path in physical space. (b) A collision-free trajectory in configuration space obtained by using global heuristics based on the path in (a).

of the find-path problem tackled. From these considerations, it follows that global heuristics are more suitable than local ones when complex paths are sought in cluttered environments, while local heuristics are more adequate than global ones when dealing with simple path finding problems.

4. Conclusions

Concerning the practical use of the algorithms described in this paper, we conclude that algebraic approaches may often be useful in problems involving at most 3 degrees of freedom of motion. This includes problems having originally more such degrees of freedom, but in which clearances are big enough to permit eliminating some such degrees (through approximation of the robot by a sphere, for instance), as well as subproblems of a larger problem.

Heuristic approaches run quicker than their algebraic counterparts in general, although they can be much slower for particularly "bad-behaved" problem instances. In these cases, where there is only one solution in a very large search space, recourse to an algebraic algorithm is mandatory.

We think that the decomposition of the motion planning problem proposed in (Ilari 1987; Ilari and Torras 1987, 1990) suggests a promising placement of the frontier between planning and control, which we expect will lead to progress in this field. In short, a model of *physical space* would be used to *plan* a path, and then the robot would be controlled in *configuration space* to follow this path. We have argued elsewhere (Torras 1990) that algebraic and heuristic approaches seem adequate for the planning stage in this decomposition, but that massively-parallel techniques should be developed for the control stage. Again, for problems with extremely small tolerances (as is the case of assembly motion planning), recourse to an algebraic algorithm may also be required.

Acknowledgement

This work has been supported in part by the Comisión Interministerial de Ciencia y Tecnología (CICYT), under the project TIC-88-0197.

References

Avnaim F. and Boissonnat J.D. (1988): "Polygonal placement under translation and rotation". *Proc. of STACS*, edited by R. Cori and M. Wirsing, *Lecture Notes in Computer Science* **294**, pp. 322-333.

Avnaim F., Boissonnat J.D. and Faverjon B. (1988): "A practical motion planning algorithm for polygonal objects amidst polygonal obstacles". *IEEE Conf. on Robotics and Automation*, Philadelphia, pp. 1656-1661.

Bajaj Ch. and Kim M-S. (1986): "Generation of configuration space obstacles. II - The case of a moving algebraic surfaces". *Tech. Rep. CSD-TR-586*, Purdue University.

Bajaj Ch. and Kim M-S. (1987): "Generation of configuration space obstacles: The case of moving algebraic curves". *IEEE Int. Conf. on Robotics and Automation*, pp. 979-984.

Bajaj Ch. and Kim M-S. (1988): "Generation of configuration space obstacles: The case of a moving sphere". *IEEE Journal of Robotics and Automation* **4(1)**, February, pp. 94-99.

Barraquand J. and Latombe J-C. (1991): "Robot motion planning: a distributed representation approach". *The Intl. J. of Robotics Research.*

Brooks R.A. (1983a): "Solving the find-path problem by good representations of free space". *IEEE Transactions on Systems, Man, and Cybernetics* **13**, pp. 190-197.

Brooks R.A. (1983b): "Planning collision-free motions for pick-and-place operations". *The Intl. J. of Robotics Research* **2(4)**, winter, pp. 19-44.

Brooks R.A. and Lozano-Pérez T. (1985): "A subdivision algorithm in configuration space for findpath with rotation". *IEEE Transactions on Systems, Man, and Cybernetics* **15**, pp. 224-233.

Canny J.F. (1988): *"The Complexity of Robot Motion Planning"*. MIT Press.

Canny J.F. and Donald B.R. (1987): "Simplified Voronoi diagrams". *Proc. ACM Symposium on Computational Geometry*, Waterloo, June.

Chatila R. (1982): "Path planning and environment learning in a mobile robot system". *Proc. of the 5th European Conference on Artificial Intelligence (ECAI)*, Orsay, July.

Donald B.R. (1984): "Global and local techniques for motion planning". *Master Thesis*, MIT, Cambridge, Massachusetts.

Donald B.R. (1987): "A search algorithm for robot motion planning with six degrees of freedom". *Artificial Intelligence* **31(3)**, pp. 295-353.

Faverjon D. (1984): "Obstacle avoidance using an octree in the configuration space of a manipulator". *Proc. IEEE Conf. on Robotics and Automation*, Atlanta, March.

Gouzènes L. (1983): "Generation of collision-free trajectories for mobile and manipulator robots". *Technical report*, LAAS du CNRS, Toulouse.

Ilari J. (1987): "Study of new heuristics to compute collision-free paths of rigid bodies in a 2D universe". *Ph.D. Thesis*, Universitat Politècnica de Catalunya, April.

Ilari J. and Torras C. (1987): "The classical 2D find-path problem: improving search efficiency by using orientation heuristics". *Proc. SPIE: Symposium on Advances in Intelligent Robotic Systems*, pp. 248-254.

Ilari J. and Torras C. (1990): "2D path planning: a configuration space heuristic approach". *The Intl. Journal of Robotics Research* **9**, pp. 75-91.

Khatib O. (1986): "Real time obstacle avoidance for manipulators and mobile robots". *The Intl. Journal of Robotics Research* **5**, pp. 90-98.

Laumond J-P. (1987): "Finding collision-free smooth trajectories for a non-holonomic mobile robot". *Proc. 10th Intl. Joint Conference on Artificial Intelligence (IJCAI)*, Milan.

Laumond J-P., Simeon T., Chatila R. and Giralt G. (1988): "Trajectory planning and motion control for mobile robots". *IUTAM/IFAC Symp. Dynamics of Controlled Mechanical Systems*, Zürich, June, Springer-Verlag.

Lozano-Pérez T. (1983): "Spatial planning: a configuration space approach". *IEEE Trans. on Computers* **32(2)**, February, pp. 108-120.

Lozano-Pérez T. and Wesley M. (1979): "An algorithm for planning collision-free paths among polyhedral obstacles". *Communications of the ACM* **22(10)**, October, pp. 560-570.

Luh J.Y.S. and Campbell C.E. (1981): "Minimum distance collision-free path planning for industrial robots with a prismatic joint". *IEEE Trans. on Automatic Control* **29(8)**, pp. 675-680.

Ó'Dúnlaing C. and Yap Ch.K. (1985): "A 'retraction' method for planning the motion of a disc". *Journal of Algorithms* **6**, pp. 104-111.

Ó'Dúnlaing C., Sharir M. and Yap Ch.K. (1983): "Retraction: a new approach to motion planning". *Proc. 15th ACM STOC*, pp. 207-220.

Ó'Dúnlaing C., Sharir M. and Yap Ch.K. (1987): "Generalized Voronoi diagrams for a ladder: II - Efficient construction of the diagram". *Algorithmica* **2**, pp. 27-59.

Schwartz J.T. and Sharir M. (1983): "On the 'piano movers' problem. II. General techniques for computing topological properties of real algebraic manifolds". *Advances in Applied Mathematics* **4**, pp. 298-351.

Torras C. (1989): "Planning for problem solving: a survey". In *"AI and Expert Systems in Scientific Computing"*, edited by R.M. Huber, C. Kulikowski, J.M. David and J.P. Krivine, IMACS Transactions Series, Baltzer Scientific Publishing Co.

Torras C. (1990): "Motion planning and control: Symbolic and neural levels of computation". *Proc. 3rd. COGNITIVA*, pp. 207-218, Madrid, November.

Udupa S.M. (1977): "Collision detection and avoidance in computer controlled manipulators". *Proc. IJCAI*, Cambridge, Massachusetts, pp. 737-748.

Whitesides S.H. (1985): "Computational Geometry and motion planning". In *"Computational Geometry"*, edited by G. Toussaint, Machine Intelligence and Pattern Recognition Series, Vol. 2, North-Holland, Amsterdam New-York Oxford.

Yap Ch.K. (1987): "Algorithmic motion planning". In *"Advances in Robotics, Vol I. Algorithmic and Geometric Aspects of Robotics"*, edited by J.T. Schwartz and Ch.K. Yap, Lawrence Erlbaum Associates.

AUTONOMOUS MOBILE ROBOTS AND TELEOPERATION

JOSÉ DEL R. MILLÁN[1], CARME TORRAS[2], MARC BECQUET[1]

[1] *Institute for System Engineering and Informatics*
Commission of the European Communities. Joint Research Centre
TP 361. 21020 ISPRA (VA). ITALY
e-mails: j_millan@jrc.it, m_becquet@jrc.it

[2] *Institut de Cibernètica (CSIC-UPC)*
Diagonal, 647. 08028 BARCELONA. SPAIN
e-mail: torras@ic.upc.es

ABSTRACT. In todays teleoperation activity, most mobile robots are completely controlled by human operators. Recent improvements in automatic control and related fields make autonomous mobile robots more feasible. Parallel distributed information processing is required to reach such an objective. In this chapter, an original technique for path finding is presented. System architecture, learning algorithm and generalization capabilities are considered. Experimental results are discussed and perspectives are addressed.

1. Introduction

Basic teleoperated mobile robots are completely controlled by human operators. In order these robots to work, operators must make even the simplest decisions —except simple inverse kinematics tasks. As a result, almost all operators' task is routine since most of the time they do not need to think for acting. Except at the beginning of the task, rarely do operators stop and plan their next actions.

This observation suggests a need of *intelligent automation*. In teleoperation, intelligent automation means that the tasks are carried out automatically with a supervisory control by human operators. A step in this direction is to develop path-planners. Operators only need to supervise that the robot follow the precomputed path. In case of failure, either the planner is invoked again or the operator takes the robot to the goal manually.

However, planning cannot be used in cases where the environment is (partially) unknown or dynamic and when it is necessary to take into account a continuous set of robot configurations and robot actions. Moreover, a robot navigating in a real environment must be able to cope with unexpected events.

In this chapter we propose a novel approach to building autonomous mobile robots. In Section 2, we put forward that an autonomous mobile robot should consist of two

M. C. Becquet (ed.), Teleoperation: Numerical Simulation and Experimental Validation, 41–53.

modules. The first of these modules is a planner. The second one is an adaptive reactive system. In Section 3 we review previous ways of implementing adaptive reactive systems. All of them rely on connectionist techniques. Section 4 is devoted to the description of our own implementation. Simulation results are given, including robustness aspects and generalization capabilities. Finally, in Section 5, some conclusions about this work are provided.

2. Path Finding as a Two-Steps Process

In (Torras, 1990; Millán & Torras, 1991a), we have argued that the *robot path finding* problem should be solved in two different stages. Firstly, a *coarse-level path* is computed in *physical space* —i.e. disregarding both the shape and the kinematics of the robot. Secondly, this path is refined to obtain a *continuous, obstacle-avoiding trajectory* in *configuration space* —which, roughly speaking, is the space of degrees of freedom of motion.

The above two-stage decomposition of the path finding problem suggests a possible separation between the path-planning and motion-control aspects of the problem. In *path planning*, the environment is considered at a coarse enough level of detail to be assumed fixed and known a priori —through access to a map— and the goal is to determine, prior to its execution, a sequence of displacements in physical space for reaching a given place. Once a path through physical space is planned at this coarse level, a *high-level motion control* problem arises since commands must be sent to the different motors to make the robot follow the path previously computed, while transforming a discrete path in physical space into a continuous trajectory in configuration space that avoids collisions with the obstacles in the way.

Having decomposed the robot path finding problem into a two-stage process, it comes out that *planning* methods are appropriate for handling the first stage but are not suitable for dealing with the second stage. The part of the problem tackled at this second stage is characterized by (*i*) a continuous set of robot configurations and of robot actions, (*ii*) a partially unknown and dynamic environment, and (*iii*) a need for coping with unexpected events —derived or not from the robot actions— and for making real-time decisions.

An alternative to handle this second stage of processing is *to build a mapping from perceived situations to correct actions, and iterate this mapping until the goal is reached.* Systems that use this kind of *situation-action rules* are known as *reactive* systems.

Reactive systems normally rely on knowledge-based techniques. However, most of them are not adaptive, i.e. they do not learn (Agre & Chapman, 1987; Arkins, 1987; Schoppers, 1987; Blythe & Mitchell, 1989). In some systems, the situation-action rules are preprogrammed explicitly by their designers (Agre & Chapman, 1987; Arkins, 1987). In other systems, the rules result from a compilation of previous planning activities (Schoppers, 1987; Blythe & Mitchell, 1989). In any case, the main limitation of the knowledge-based systems developed up to now is the need to specify actions to be taken in all possible situations the agent may find itself in. On the other hand, the knowledge-based reactive systems able to construct by themselves an internal model of the environment have only been applied to simple tasks (Rivest & Schapire, 1987). Mozer & Bachrach (1989) have implemented the Rivest & Schapire's system in a connectionist network, which has been shown to perform better than the original symbolic version in several ways.

Brooks' approach (1986) is a novel way of building reactive systems. Its originality is based on the subsumption architecture and also on hard-wiring sensors to effectors rather than representing the rules. This approach, however, does not overcome the above-mentioned limitation. Nevertheless, it emphasizes a key issue: reactive behavior should involve parallel distributed processing.

This chapter focusses on the second stage of the robot path finding problem. In particular, we propose to develop adaptive reactive systems by using *connectionist learning techniques*. For a review of planning approaches, see the chapter by Torras. Section 5 provides some more information on how to combine the two stages.

3. Connectionist Approaches to Robot Path Finding

Subsymbolic systems —connectionist or not— process information in a massively-parallel distributed manner. Previous subsymbolic attempts to tackling the robot path finding problem fall into two categories: constructive and pattern-matching systems.

Pattern-matching connectionist path-finders are *adaptive* reactive systems, where the mapping is learned as paths are generated (Massone & Bizzi, 1989; Nguyen & Widrow, 1989; Sutton, 1990). In addition, they do not need to codify explicitly all the situation-action rules since they exhibit strong generalization capabilities.

Constructive connectionist path-finders lie somewhere in between path-planners and pattern-matching connectionist path-finders systems (Jorgensen, 1987; Graf & LaLonde, 1988; Steels, 1988; Mel, 1989). They partially use planning but they do not assume as many constraints about the environment as path-planners do.

Of all the systems mentioned in this section, only the one of Steels does not properly constitute a connectionist approach.

The next two subsections briefly outline these previous connectionist approaches to robot path finding. For a more detailed review of them —and others—, together with an assessment of their merits and limitations, see (Millán & Torras, 1991a).

3.1. CONSTRUCTIVE PATH-FINDERS

These systems codify the environment through connectionist techniques so as to permit the exploration of configuration space in parallel and at a low computational cost. Constructive path-finders, however, suffer to some extent from the inherent limitations of planning.

The configuration space on which the search is carried out can be constructed either through learning or not. Because *pure constructive* path-finders (Steels, 1988) do not require a learning phase, they generate the configuration space using a few computational resources. Nevertheless, they need a complete model of the robot's environment. *Mixed constructive* path-finders (Jorgensen, 1987; Graf & LaLonde, 1988; Mel, 1989), on the other hand, exhibit the opposite advantages and drawbacks: the learning phase may consume a lot of time and memory, but complex relationships between the states can be captured.

3.2. PATTERN-MATCHING PATH-FINDERS

In the case of pattern-matching path-finders, how can the sensory-motor transformation rules be learned? If sample paths expressed in *motor space* are available, then it is possible to build the mappings through *supervised* learning (Massone & Bizzi, 1989). The main difficulty in this case is to obtain the training data, i.e. the desired motor outputs. In certain situations, however, one can be provided only with sample paths expressed in *configuration space.* The *system identification approach* (Nguyen & Widrow, 1989) has arisen to handle this case. It uses supervised learning, but it requires a model of the robot to be controlled.

A more generally applicable possibility is to discover those rules by a trial-and-error procedure which evaluates —according to a given criterion— each sensory-motor transformation in order to obtain a scalar *reinforcement* signal (Sutton, 1990). Due to the kind of training information used, reinforcement systems theoretically require more learning time than supervised systems: in the latter, each unit is provided with specific information to adapt its behavior while, in the former, the same information is used by all the units.

This theoretical limitation of reinforcement systems, however, is not strongly supported by experimental data. On the contrary, the (relatively) simple experiments conducted by Millán & Torras (1990) and Sutton (1990) on path finding show that using a reinforcement method does not imply a time-consuming learning phase. Even more, it should be noted that using reinforcement systems is preferable when there are many possible satisfactory actions for a given stimulus. In such cases, it is more appropriate to let the system learn its own mapping rather than to impose an arbitrary one.

Despite the fact that these reinforcement systems have not yet succeeded in dealing adequately with all the above-mentioned characteristics of the problem, we believe that they are the most suitable for tackling the second stage of processing. Our bet is supported by two facts. First, the only connectionist path-finders developed so far that are able to avoid obstacles while reacting in real time are based on reinforcement learning. Second, reinforcement systems do not require as many resources as supervised systems do —i.e. training data and/or previous model of the robot— and in any case they can be combined with supervised methods to improve their behavior. Two immediate places for incorporating supervised methods in a reinforcement system are the *critic* element (Barto et al., 1983; Sutton, 1984) and the *learning rule used for updating the hidden layer of the path-finder* (Anderson, 1986).

4. A Reinforcement Connectionist Path-Finder

The particular instance of the robot path finding problem we are interested in is characterized by (*i*) a continuous set of robot configurations and of robot actions, (*ii*) a partially unknown and dynamic environment, and (*iii*) a need of coping with unexpected events and of making real-time decisions. This section is devoted to present a *reinforcement connectionist system* able to find and learn the suitable situation-action rules so as to generate feasible paths for a mobile robot in a 2D environment, while appropriately dealing with the three problem characteristics above. The criterion used for evaluating the quality of a path is a compromise between minimizing path length and maximizing the distance to the obstacles.

An intelligent agent interacting with its environment is said to improve when it learns to react with a more and more useful action —where *usefulness* is defined in terms of the agent's goal— to each situation. The *associative reinforcement learning (ARL)* problem (Barto et al., 1981; Sutton, 1984; Barto & Anandan, 1985) offers a simple and general framework for developing this kind of systems. Simply stated, the ARL problem is that of learning to associate with each stimulus **X** the action **Y** that maximizes reinforcement z —either present, future or cumulative.

The approach to the ARL problem adopted here is *to estimate the gradient of the reinforcement with regard to the system's actions by measuring the correlation between variations in actions and variations in reinforcement.* Once this gradient is available, it is used for modifying the agent. This learning paradigm is known as *reinforcement learning.*

4.1. CODIFICATION SCHEME

The *input* to the path-finder consists of an *attraction force* and several *repulsion forces* exerted respectively by the goal configuration and the obstacles on the current configuration. The robot does perceive only obstacles in its immediate surroundings.

Let the *shortest path line (SPL)* and the *shortest path vector (SPV)* be, respectively, the line and the vector that connect the robot current and goal configurations. The direction of the attraction force is that of the SPV, its intensity being an inverse exponential function of the distance between the current and goal configurations.

Each repulsion force represents the resistance of an obstacle toward the robot SPV following process. Each such force has the direction of a bisector of the SPL and its perpendicular, starting at the current configuration and heading the opposite quadrant to that where the obstacle lies. Because the directions of the repulsion forces are specified with respect to the direction of the attraction force, they are implicit in the codification and therefore are not included in the input to the system. The intensity of each repulsion force depends on three factors. The first factor is aimed at avoiding obstacles in the proximity of the SPV. The second allows to avoid obstacles near to the robot current configuration. The third acts in case the SPV intersects an obstacle: the deeper the penetration into the obstacle the more distant from the SPV will be the next robot motion.

The SPL and its perpendicular divide the workspace into four quadrants. Repulsion forces in a quadrant try to deflect the robot move toward the center of the opposite quadrant. Thus all the information related to the repulsion forces can be reduced to four signals: *intensity of the environmental repulsion from each quadrant.*

In short, *the number of input signals* to the path-finder *is independent of the environment*, and these signals are five: the intensity of the attraction force and the four environmental repulsion intensities. Since the output of the path-finder is computed with respect to the direction of the attraction force (see below), this direction does not need to be included in the input of the system. Each input signal is codified as a real number in $[0, 1]$.

The *output* of the path-finder represents the step taken by the robot and is codified as a move in *relative cartesian coordinates* with regard to the SPV. That is, the axis X is the SPL. Both coordinates take real values in $[-1, 1]$. The maximum distance the robot can cover in a single step is limited by the *perception range*: the robot could collide with obstacles "not completely" perceived since the robot concentrates on its immediate neighbourhood.

The *reinforcement signal* is a measure of how good is the answer of the system to a particular stimulus. It is calculated on the basis of the quality of the configuration reached by the robot —a combination of the attraction and repulsion factors— and the way in which this configuration has been reached. It is a real value in $[-1, 1]$.

The determination of both the input and goodness degree is reminiscent of the *potential field approach* to path planning (Khatib, 1986).

Two are the benefits of this codification scheme. First, since the goal is codified in the input information and the input is independent of the environment used during the learning phase, it allows to transfer the knowledge acquired for a situation to a different one. Second, it reduces the complexity of the task to be solved. In the robot path finding problem, the consequences of an action can emerge later in time. Thus, actions must be selected based on both their short- and long-term consequences. Since the environmental reinforcement signal is computed using *global information* —i.e. it is based not on the current robot configuration but on the SPV—, the path-finder gets a measure of the short- and long-term consequences of an action only one time-step after it has been executed. The task is reduced to learn, for each stimulus, the action which maximizes the environmental reinforcement signal.

4.2. SYSTEM ARCHITECTURE

The path-finder is made of two elements, namely the *step generator* and the *critic*, and interacts with its environment as depicted in Figure 1. At each time t, the environment provides the path-finder with the input pattern $\mathbf{X}(t) = (x_1(t), x_2(t), x_3(t), x_4(t), x_5(t))$, together with the *environmental reinforcement signal* $z(t)$. The input pattern is fed to both the step generator and the critic. Nevertheless, the step generator does not receive directly the environmental reinforcement signal but the *heuristic reinforcement signal* $h(t)$ elaborated by the critic. The latter is an enrichment of the former based on past experience of the path-finder when interacting with the environment. The step generator produces instantaneously an output pattern $\mathbf{Y}(t) = (y_1(t), y_2(t))$ that it is the output of the path-finder. The environment receives this action $\mathbf{Y}(t)$ and, at time $t+1$, sends to the path-finder both an evaluation $z(t+1)$ of the appropriateness of the action $\mathbf{Y}(t)$ for the stimulus $\mathbf{X}(t)$ and a new stimulus $\mathbf{X}(t+1)$.

To illustrate the usefulness of the critic, let us consider the two following situations. First, the output of the step generator at time t $\mathbf{Y}(t)$ is the best action possible for a given stimulus $\mathbf{X}(t)$, but the associated environmental reinforcement $z(t+1)$ is -0.5. For example, if the current and goal configurations are located at opposite sides of an obstacle, then the best move could be to send the robot away from the goal. Alternatively, the step generator receives an environmental reinforcement signal $z(t+1)$ of $+0.5$ as a consequence of producing the action $\mathbf{Y}(t)$ for the stimulus $\mathbf{X}(t)$, when it could have generated a better action $\mathbf{Y}'(t)$ which allows to reach the goal configuration from the current one. These two situations show that the environmental reinforcement signal is not sufficient. To solve the path finding problem, what the step generator needs is high reinforcement in the first case, and low, perhaps negative, reinforcement in the second case. The goal of the critic is to transform the environmental reinforcement signal into a more informative signal. The resultant signal is the heuristic reinforcement signal.

4.3. LEARNING ALGORITHM

Among the different kinds of reinforcement learning algorithms proposed in the literature, we adopt the *associative search, AS,* algorithm (Barto et al., 1981; Sutton, 1984; Williams, 1987) because it is more appropriate for the specifications of the output and reinforcement signals we have.

A central issue for any reinforcement system is to explore alternative actions for the same stimulus. *Stochastic units* provide this source of variation. But the stochastic behavior of the path-finder should tend to be *deterministic* with learning. Otherwise, the path-finder could not generate *stable solution paths* after it eventually discovers them. By using a suitable algorithm for transforming the stochastic units into deterministic ones (Millán & Torras, 1990) the path-finder not only discovers the suitable situation-action rules, but it also does it in a very short time.

4.3.1. The Basic AS Algorithm. Continuous stochastic units compute their output in three steps (Gullapalli, 1988), as expressed in Equations (1) through (5). The only difference with respect to Gullapalli's formulation is the way in which the variance is computed.

Since the signals we are interested in are continuous, a separate control of the location being sought (*mean*) and the breadth of the search around that location (*variance*) is needed. The first step is to determine the value of these parameters. The mean μ should be an estimation of the optimal output. A simple way is to let $\mu_i(t)$ equal a weighted sum of the inputs $s_j(t)$ to the unit i plus a threshold $\theta_i(t)$:

$$\mu_i(t) = \sum_{j=1}^{n} \left(w_{ij}(t)s_j(t)\right) + \theta_i(t). \tag{1}$$

The standard deviation σ should be small if the expected output of the step generator is close to the optimal, and it should be high in the opposite case. Since the heuristic reinforcement signal provides this comparative measure of the output goodness, $\sigma(t)$ should depend on the *expected heuristic reinforcement*, $\widehat{h}(t)$:

$$\sigma(t) = k_\sigma * \widehat{h}(t), \tag{2}$$

where k_σ is a constant and $\widehat{h}(t)$ is a trace of the absolute value of past heuristic reinforcement received:

$$\widehat{h}(t) = \xi * \text{abs}\left(h(t)\right) + [1 - \xi]\widehat{h}(t-1), \tag{3}$$

with ξ being a constant in $[0, 1]$.

As a second step, the unit calculates its *activation level* $a_i(t)$ which is a *normally distributed random variable*:

$$a_i(t) = N\left(\mu_i(t), \sigma(t)\right). \tag{4}$$

Finally, the unit computes its output $s_i(t)$:

$$s_i(t) = f\left(a_i(t)\right) = \frac{2}{1 + e^{-\beta a_i(t)}} - 1, \tag{5}$$

where β is a constant in $[0,1]$.

A deterministic unit computes its output as a weighted sum of its input:

$$s_i(t) = f\left(\sum_{j=1}^{n}\left(w_{ij}(t)s_j(t)\right) + \theta_i(t)\right), \tag{6}$$

where $f(\cdot)$ is the same function as in Equation (5).

In the AS family of algorithms, the weights are modified according to the following general expression:

$$\Delta w_{ij}(t) = \alpha h(t) e_{ij}(t-1), \tag{7}$$

where α is the *learning rate* and e_{ij} is the *eligibility factor* of w_{ij}. The eligibility factor of a given weight is a measure of how influential that weight was in choosing the action. Each particular version of this general algorithm differs from the others in the way h and e_{ij} are calculated.

In the experiments reported here, we use the version of the AS algorithm that best suits the robot path finding problem. This version has been identified after a comparative study on twenty five versions of the basic AS algorithm:

$$h(t) = z(t) - \widehat{z}(t-1), \tag{8}$$

$$e_{ij}(t) = \frac{\partial \ln N}{\partial w_{ij}}(t) = s_j(t)\frac{a_i(t) - \mu_i(t)}{\sigma^2(t)}, \tag{9a}$$

$$e_{ij}(t) = s_i(t) * s_j(t), \tag{9b}$$

where $\widehat{z}$ is the *expected primary reinforcement* calculated by the critic. In order to undertake this prediction task, the critic is built as a second network out of deterministic units, and since it is provided with input/output pairs, a *supervised algorithm* is used. e_{ij} is calculated in two different manners according to the kind of units of the step generator. If the units are stochastic, it is computed in such a manner that the learning rule corresponds to a *gradient ascent* mechanism on the expected environmental reinforcement (Williams, 1987). If the units are deterministic, it is the *Hebbian rule.*

4.3.2. A Strategy for Discovering Stable (Quasi) Optimal Paths. It has been stated above that in order to obtain stable solution paths, stochastic units should become deterministic as learning proceeds. The way in which σ is computed guarantees that the breadth of search will diminish asymptotically to zero as the path-finder learns. But this is not acceptable to solve the problem efficiently.

A mechanism for accelerating the generation of stable solution paths is the following. When the path-finder discovers, after a certain experience with the environment, *acceptable* paths, it might be more suitable to search the solution paths in a more "controlled" manner by *transforming the stochastic units of the step generator into deterministic units.* In addition, the weights are not updated after each step, but after the path has been generated and only if the path is not *(quasi) optimal, qo-path.* The acceptability and optimality criteria are defined by the boolean expressions:

$$\begin{aligned} qo\_path = {} & \left(length < k_{qo_{length}} * \mathrm{dist}[conf_{start}, conf_{goal}]\right) \\ & \wedge \left(minimum\_clearance < k_{qo_{clear}}\right), \end{aligned} \tag{10}$$

$$\begin{aligned} acc\_path = {} & \left(length < k_{acc_{length}} * \mathrm{dist}[conf_{start}, conf_{goal}]\right) \\ & \wedge \left(minimum\_clearance < k_{acc_{clear}}\right), \end{aligned} \tag{11}$$

where $k_{qo_{length}}$, $k_{qo_{clear}}$, $k_{acc_{length}}$, and $k_{acc_{clear}}$ are workspace-dependent constants.

The weight updates are delayed and not applied when a qo-path is generated because, otherwise, the changes to the weights could probably prevent the path-finder from reproducing it. Since the output signals are continuous and the optimality criterion is very severe, even little weight modifications could alter the step generator as to produce actions sufficiently away from the optimal ones.

The process for discovering a stable qo-path requires a further refinement. The fact of obtaining an acceptable path does not mean that the qo-path is nearby. The discovered acceptable path could be close to a local optimum. Even more, if collisions happen while deterministic units are used, the acceptable paths could be lost because of the penalty received. For these reasons, if after consuming a fixed quantity of computational resources the path-finder does not discover a qo-path, then the deterministic units should turn back to being stochastic again. Another acceptable solution should then be obtained. Figure 2 shows the final algorithm for discovering a stable qo-path.

4.4. EXPERIMENTAL RESULTS

The system has been implemented in Common Lisp on a VAX station 2000. The *step-generator* we have used to carry out the simulations has three layers of units, all the units in a layer being connected to all the units in all the layers above. The hidden layer is made of four units. As usual, the functionality of an input layer consists simply in forwarding the signal it receives. The hidden and output units of the step-generator are stochastic units that become deterministic as learning proceeds. The *critic* has a hidden layer of four deterministic units and, obviously, one output deterministic unit.

In the simulations, the robot moves inside the workspace until either it reaches the goal or it collides. The learning phase is carried out in an incremental way. Initially, the path-finder has to learn to generate a feasible path from the first starting configuration in the training set. Then, it tries to generate a feasible path from the second starting configuration. As the necessary knowledge required for solving this new situation is discovered and codified, part of the previous knowledge could be lost. So, the path-finder must learn again to generate a feasible path from the first starting configuration. The iteration "learn to generate a feasible path from the new starting configuration—recover the ability to generate feasible paths from the previous starting configurations" is repeated until the path-finder is able to generate feasible paths from all the starting configurations.

In (Millán & Becquet, 1991; Millán & Torras, 1991b) we demonstrate the feasibility of our approach. The success of the path-finder lies in two key issues: the codification scheme adopted and the algorithm used to discover stable solution paths. They not only lead to very quick learning, but are also responsible for two additional positive features, namely its noise tolerance and its generalization capabilities. Some simulation results follow.

Figure 3 shows the behavior of the path-finder once the learning phase is finished. Panel A depicts the paths generated by the path-finder from every starting configuration in the training set. Obstacles are shown as circles and every initial configuration and the goal are respectively represented by a triangle and a square. In panel B, instances of the situation-action rules discovered by the path-finder are illustrated; for every starting configuration considered —little circles— the move to be taken by the robot is shown. The path-finder is able to produce stable collision-free paths from almost all the initial configurations in the training set. The number of steps required to reach this state of the path-finder has been 77315. The only situations not properly handled by the path-finder are a subset of the most difficult ones, that is those in which an obstacle lies in between the goal and the current robot configuration and is very close to this configuration. These situations may be handled by getting appropriate guidance from a symbolic path-planner (Millán & Torras, 1991b).

Until now, we have assumed that the robot can perfectly perceive the workspace. Nevertheless, a robot navigating in a real environment is subject to noisy and inaccurate measurements. To test the ability of our path-finder to work under real conditions, a 20% of white noise is added to each input signal. Figure 4 depicts the resulting behavior. Comparing this figure with Figure 3, panel B, it is evident that the path-finder demonstrates a *strong robustness*, since both "maps" are very similar.

Figure 3, panel B, illustrates some of the generalization abilities of the path-finder. It tackles many more situations than those perceived during the learning phase. In addition, the path-finder is also able to navigate in workspaces different from that used during learning. Figure 5 shows the behavior of the path-finder when both the goal and the number and location of the obstacles have changed. The result of this experiment proves that the *situation-action rules learned are workspace-independent.*

5. Conclusions

Different chapters of this book highlight the importance of intelligent automation in the field of teleoperation. In this chapter, we have dealt with how to build autonomous mobile robots to be used in teleoperated environments. We have also presented an approach to do that.

In Section 2 we put forward that the robot path finding problem should be solved in two differen stages. We also stressed that planning methods are not suitable for dealing with the second stage of processing and argued that this stage should be carried out by means of adaptive reactive systems. In Section 3 we identified reinforcement-based pattern-matching connectionist path-finders as the most suitable systems for handling this part of the problem, and constructive connectionist path-finders as a kind of systems lying in between planning path-finders and pattern-matching path-finders. We also reviewed these two kinds of connectionist path-finders. In Section 4 we presented a particular connectionist path-finder and provided results demonstrating the feasibility of our approach.

The formulation of the problem used to test this approach is a difficult one, since the input and output are continuous, the environment is partially unknown, and the optimality criterion is severe. The problem, however, has been simplified by assuming a point robot and circular obstacles. To overcome these limitations, we are currently extending the prototype for dealing with a 2D mobile robot and polygonal obstacles.

Our hypothesis of departure is that the robot path finding problem could be solved efficiently if symbolic planning were interfaced to subsymbolic obstacle avoidance. Our current research is oriented towards designing a *hybrid path-finder* by coupling a geometric global planning approach such as that in (Ilari & Torras, 1990) with the connectionist local obstacle avoidance approach described in this chapter.

In (Millán & Torras, 1991b) we illustrate with a very simple example the potential benefits of integrating symbolic and subsymbolic techniques according to this general framework. A symbolic path-planner suggests intermediate configurations —subgoals or

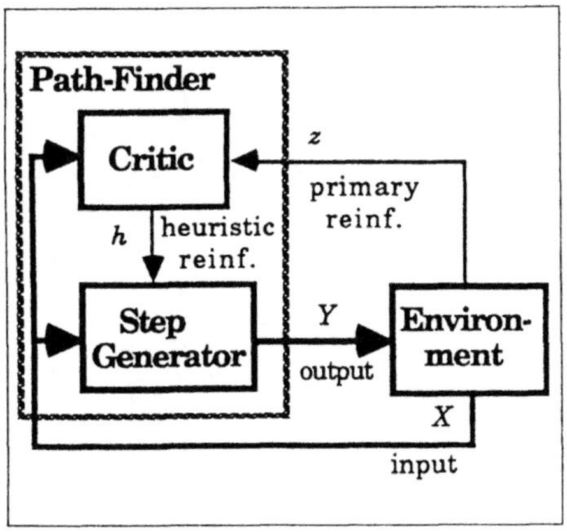

Figure 1. Connectionist path-finder according to the ARL architecture.

```
Algorithm discover_stable_qo-path
begin
  initializations
  repeat
    repeat
      generate_path (immediate_learning,
                     stochastic_units)
    until (acceptable_path and
           enough_experience)
    repeat
      generate_path (delayed_learning,
                     deterministic_units)
    until (qo-path or no_more_resources)
  until qo-path
end
```

Figure 2. Algorithm for discovering a stable QO-path.

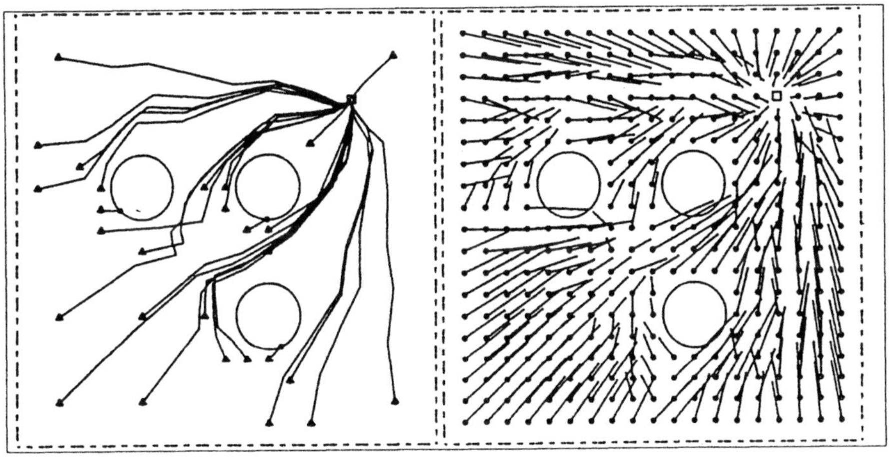

Figure 3. Behavior of the path-finder.

landmarks— that the path has to go through. The path-planner is invoked either at the beginning of the task or whenever the right course of actions seems to be wrong. This happens when the path-finder cannot find suitable actions to handle the current situation or when the robot deviates considerably from the planned physical path.

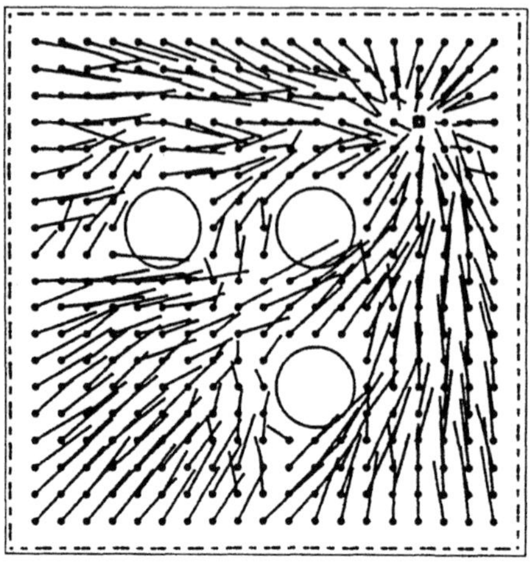

Figure 4. Noise tolerance exhibited by the path-finder.

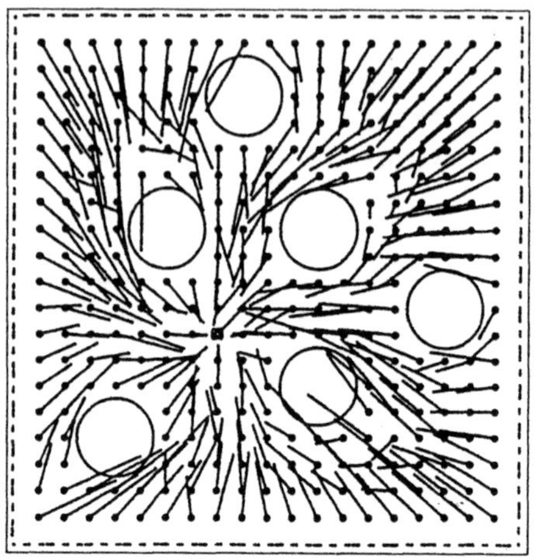

Figure 5. Generalization abilities: A new goal and more obstacles.

References

Agre, P.E., & Chapman, D. (1987). Pengi: An implementation of a theory of activity. *Proc. 7th AAAI Conference*, 268–272.

Anderson, C.W. (1986). Learning and problem solving with multilayer connectionist systems. Ph.D. Thesis, Dept. of Computer and Information Science, University of Massachusetts, Amherst.

Arkins, R.C. (1987). Motor schema based navigation for a mobile robot: An approach to programming by behavior. *Proc. IEEE International Conference on Robotics and Automation*, 264–271.

Barto, A.G., & Anandan, P. (1985). Pattern-recognizing stochastic learning automata. *IEEE Transactions on Systems, Man, and Cybernetics*, **15**, 360–374.

Barto, A.G., Sutton, R.S., & Anderson, C.W. (1983). Neuronlike elements that can solve difficult learning control problems. *IEEE Transactions on Systems, Man, and Cybernetics*, **13**, 835–846.

Barto, A.G., Sutton, R.S., & Brouwer, P.S. (1981). Associative search network: A reinforcement learning associative memory. *Biological Cybernetics*, **40**, 201–211.

Blythe, J., & Mitchell, T.M. (1989). On becoming reactive. *Proc. 6th International Workshop on Machine Learning*, 255–259.

Brooks, R.A. (1986). A robust layered control system for a mobile robot. *IEEE Journal of Robotics and Automation*, **2**, 14–23.

Graf, D.H., & LaLonde, W.R. (1988). A neural controller for collision-free movement of general robot manipulators. *Proc. IEEE 2nd International Conference on Neural Networks*, Vol I, 77–84.

Gullapalli, V. (1988). A stochastic algorithm for learning real-valued functions via reinforcement feedback. Technical Report COINS–88–91, Dept. of Computer and Information Science, University of Massachusetts, Amherst.

Ilari, J., & Torras, C. (1990). 2D path planning: A configuration space heuristic approach. *The International Journal of Robotics Research*, **9**, 75–91.

Jorgensen, C.C. (1987). Neural network representation of sensor graphs in autonomous robot navigation. *Proc. IEEE 1st International Conference on Neural Networks*, Vol IV, 507–515.

Khatib, O. (1986). Real time obstacle avoidance for manipulators and mobile robots. *The International Journal of Robotics Research*, **5**, 90–98.

Massone, L., & Bizzi, E. (1989). Generation of limb trajectories with a sequential network. *Proc. International Joint Conference on Neural Networks*, Vol II, 345–350.

Mel, B.W. (1989). MURPHY: A neurally-inspired connectionist approach to learning and performance in vision-based robot motion planning. Ph.D. Thesis, Graduate College, University of Illinois, Urbana-Champaign.

Millán, J. del R., & Becquet, M. (1991). A reinforcement connectionist path-finder. *Proc. 6th International Conference on CAD/CAM, Robotics and Factories of the Future.*

Millán, J. del R., & Torras, C. (1990). Reinforcement learning: Discovering stable solutions in the robot path finding domain. *Proc. 9th European Conference on Artificial Intelligence*, 219–221.

Millán, J. del R., & Torras, C. (1991a). Connectionist approaches to robot path finding. In O.M. Omidvar (ed.), *Progress in Neural Networks Series, Vol 3.* Norwood, NJ: Ablex.

Millán, J. del R., & Torras, C. (1991b). Learning to avoid obstacles through reinforcement. In L. Birnbaum & G. Collins (eds.) *Machine Learning: Proceedings of the 8th International Workshop*, 298–302. San Mateo, CA: Morgan Kaufmann.

Mozer, M.C., & Bachrach J. (1989). Discovering the structure of a reactive environment by exploration. Technical Report CU-CS-451-89, Dept. of Computer Science, University of Colorado, Boulder.

Nguyen, D., & Widrow, B. (1989). The truck backer-upper: An example of self-learning in neural networks. *Proc. International Joint Conference on Neural Networks*, Vol II, 357–363.

Rivest, R.L., & Schapire, R.E. (1987). A new approach to unsupervised learning in deterministic environments. *Proc. 4th International Workshop on Machine Learning*, 364–375.

Schoppers, M.J. (1987). Universal plans for reactive robots in unpredictable environments. *Proc. 10th International Joint Conference on Artificial Intelligence*, 1039–1046.

Steels, L. (1988). Steps towards common sense. *Proc. 8th European Conference on Artificial Intelligence*, 49–54.

Sutton, R.S. (1984). Temporal credit assignment in reinforcement learning. Ph.D. Thesis, Dept. of Computer and Information Science, University of Massachusetts, Amherst.

Sutton, R.S. (1990). Integrated architectures for learning, planning, and reacting based on approximating dynamic programming. *Proc. 7th International Conference on Machine Learning*, 216–224.

Torras, C. (1990). Motion planning and control: Symbolic and neural levels of computation. *Proc. 3rd COGNITIVA Conference*, 207–218.

Williams, R.J. (1987). Reinforcement-learning connectionist systems. Technical Report NU-CCS-87-3, College of Computer Science, Northeastern University, Boston.

KINEMATIC CALIBRATION IN REMOTE HANDLING AND TELEOPERATION ENVIRONMENT

J.-M. RENDERS
Institute for Systems Engineering and Informatics, TP 210,
Commission of the European Communities,
Joint Research Center, I-21020 Ispra (VA), Italy

Abstract: *Maintenance and more generally intervention in hostile environment use remote handling tools; between them, robotics technology is of prime importance, because operations are then performed without the presence of man in the loop for most actions; besides flexibility and automation of routine jobs, the use of robotics tools can improve safety aspects, by allowing the operator to focus on supervision tasks. Furthermore, off-line programming of such robotics systems is highly desirable, because other functioning mode, such as "teach-repeat" mode, are no longer feasible in hostile environment. Nevertheless, off-line programming of the teleoperation tasks requires robot's accuracy for safety and reliability reasons. This implies kinematic calibration of the robot, i.e. identification of a more accurate relationship between joint encoder readings and the actual position of the end-effector. This paper presents two techniques to achieve this goal. The fisrt one (Part I) deals with geometrical parameter identification, by using the traditional estimator theory in statistics (maximum likelihood method), while the second one (Part II) uses neural network techniques for the correction of non-geometrical errors. Experimental and simulation results show that the robot accuracy can be improved by a factor of more than 50.*

PART I: GEOMETRICAL PARAMETERS IDENTIFICATION AND ERROR COMPENSATION BY MAXIMUM LIKELIHOOD ESTIMATOR

1. Problem Statement

Maintenance and more generally intervention in hostile environment use remote handling tools; between them, robotics technology is of prime importance, because operations are then performed without the presence of man in the loop for most actions; besides flexibility and automation of routine jobs, the use of robotics tools can improve safety aspects, by allowing the operator to focus on supervision tasks .

One of the most important breakthroughs in robotics technology for remote handling is the off-line programming process in CAD system that makes possible the design and layout of remote handling tools, mission planning as well as task simulation. The most important output of this simulation is a task file that can be downloaded into the robot control system to perform the desired motions. Much work is undertaken nowadays to make off-line programming an attractive tool that will enlarge the spectrum of robot applications in manufacturing processes and to transpose this technology in hostile environments. Furthermore, off-line programming of such

M. C. Becquet (ed.), Teleoperation: Numerical Simulation and Experimental Validation, 55–86.

robotics systems is highly desirable, because other functioning modes, such as the "teach-repeat" mode - which is a convenient technique in manufacturing processes since the operator can be as close as desirable from the end-effector - are no longer feasible in hostile environment. Nevertheless, off-line programming of the teleoperation tasks requires robot's accuracy for obvious safety and reliability reasons.

The major problem is the difference between the nominal geometry of a robot, based on the robot design specifications, and the real geometry of the same robot affected by manufacturing tolerances, mounting errors during robot link assembly and kinematic model simplifications in the control unit.

Nominal models used are simple and are based on several assumptions, such as parallelism or orthogonality of the axes of the joints. Differences between nominal and real kinematic models also arise from non-geometrical errors: link and joint flexibility, backlash, gear runout, *etc.*

Robot kinematic calibration consists of identifying a more accurate geometrical relationship between the joint encoder readings and the actual position of the end-effector; then the robot positioning software is changed according to this relationship identified. Kinematic calibration involves four steps: modeling, measurement, identification and correction.

The purpose of this part of the paper is to present a new calibration method including a new algorithm to identify differences between the nominal and real robot geometry, and an original experimental set-up to apply the method in a real case. In this part, non-geometrical errors are assumed to be negligible; in the second part, a method to identify and compensate for this kind of error will be presented.

Robot position error models have been widely studied. A linearized geometrical error model has been established by Veitschegger[18]; it was used for *a priori* accuracy analysis, but not for identification and calibration purposes. The same thing can be said about the work of Menq[20]; he studied the statistical properties of the position error of the end-effector employing probabilistic characterization of the unknown geometrical parameters; but his paper does not deal with identification, measurement and calibration problems.

Hollerbach[3] and An[4] used Veitschegger's linearized error model for geometrical parameter identification and gave some experimental results. However, they did not include position and orientation errors of the measuring device in the modeling process, and, using the common least-squares method for identification, they did not take into account *a priori* information (sensor accuracy, manufacturing tolerances, *etc.*). Moreover, no significant experimental results were obtained, because of insufficient accuracy of the measuring device.

Roth[21] gives an overview of Robot Calibration, discussing modeling, measurement, identification and correction issues, but no mathematical developments or experimental results are presented. Whitney[13] uses a model including non-geometrical errors and a least squares numerical search algorithm on theodolite measurements. The error model is completely non-linear and, unfortunately, the definition of geometrical parameters to be identified (six for each link) is rather unusual and related to the measuring instruments. Judd[14] considers mainly non-geometrical errors (gear train errors, joint and link flexibility, *etc.*) and proposes error models which can be used for identification with a common least squares procedure.

The contribution of this work, in the area of kinematic calibration, is composed of two main points: firstly, experimental results in arm calibration using a new kind of motion tracking system are reported; unlike other measurement methods already proposed in the literature (Veitschegger[1], An[4], Payannet[9], Ishii[12], Whitney[13], Gilby[16]), this system is simple and economically attractive, while providing sufficient accuracy for calibration purpose. Secondly, an algorithm for estimating geometrical parameter errors is presented, taking into account the following characteristics: sensor accuracy, resolver resolution, backlash in transmission units, manufacturing tolerances and, finally, position and orientation errors of the measuring device. These limits are used for building a global error model and a maximum likelihood estimator. Such an approach, while yielding identification results with a sounder theoretical justification, has not been adopted in previous work.

This part of the paper is organized in the following way. In section 2, we introduce the definition of the geometrical parameters; the effects of non-geometrical errors and the measuring device limitations are discussed; the differential modeling with the maximum likelihood estimator and its properties are described; finally, we establish the global model, including position errors of the measuring device. Section 3 presents two experiments carried out on a robot with six degrees of freedom. Different ways to use the improved kinematic model obtained after identification are described in section 4. Finally, concluding remarks appear in section 5.

2. Error Model

In this section we will build up the global model for error identification. This model depends on the choice of parameters used to describe the robot geometry and on the possible consideration of non-geometrical errors; the error model is written in a linearized form so that it can be directly exploited in the identification process. A maximum likelihood estimator is built, taking into account possible backlash in transmission units, measuring device limits and manufacturing tolerances.

2.1. Geometrical parameters

A classical choice for the geometrical parameters used to describe the robot geometry is the Denavit and Hartenberg parameters $(\theta_i, d_i, \alpha_i, a_i)$. But it is well known that, with this set of four parameters, small geometrical errors do not lead to small variations of the parameters. For example if two successive rotational joints are parallel, the common normal defining the distance between those two axes may be arbitrarily located. If they are slightly non parallel, this distance may greatly vary in magnitude and position.

For this reason, it can be convenient to use an extra parameter (Veitschegger[18]): this extra parameter defines a rotation of a frame (i) to a new frame (i') resulting from non-exactly parallel rotational axis, around the y_i axis: it is called the link twist β_i (see fig. 1). Without this link twist, the non-parallelism must be compensated for by artificial modification to the values of the length a_i and the offset d_i of the link, even if they were initially correct.

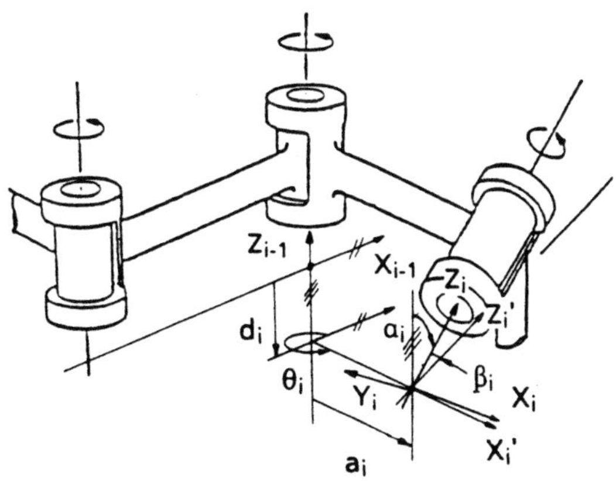

Fig.1. Definition of the Denavit and Hartenberg parameters, including the link twist β_i

This twist angle β_i is useful only for successive nominally parallel rotational joint axis. In this case, it is substituted for the joint offset error d_i, so that we are allowed to fix the dependent translation d_i and to consider that it is error free. In other cases it is set to zero.

Nominal values of those geometrical parameters can be different from the real ones in the robot system because manufacturing tolerances and assembly tolerances affect the kinematic model of the robot; moreover, the electrical zeroes of the joint encoders do not generally coincide with the mechanical zeroes of the joints themselves.

In section 2.4 and subsequent sections, we develop an identification algorithm to find the errors for the geometrical parameters.

2.2. Non-geometrical errors

As depicted in section 2.1, geometrical parameters can be affected by errors on their nominal values. But other error sources are present in a robot structure. We will call them non-geometrical errors sources.

Among these, the error sources which have the most significant effect on accuracy are:
- joint flexibility
- link flexibility
- gear transmission error
- backlash in gear transmission
- temperature effect

The second part of the paper will examine this kind of errors in detail.

2.3. Measuring device limitations

The measurement of a 3-dimensional (3D) trajectory with high accuracy, large bandwidth and large displacement is still difficult to achieve. To illustrate this statement we will briefly discuss five major means to measure 3D trajectories of the end effector.

String operated potentiometric transducers may be used to compute the position of a specific point of the manipulator on which three strings are connected. This system can be accurate enough and large displacements are allowed, but its bandwidth is limited and the system reliability must be carefully considered. Such a system is presented in Payannet[9] with experimental results.

Laser measurement systems are highly accurate, their bandwidths are sufficient and large displacements are allowed, but currently they can be used only for linear motions. Various research is in progress to develop 3D measurements laser systems, but motion speed seems to be limited (see Gilby[16]).

Ultrasonic 3D measurements systems are available on the market. Along each cartesian axis the resolution reaches 0.1 mm and sampling rate is about 50 Hz. An important disadvantage of this system is the sensitivity of sound speed in the air to temperature and humidity. Thus calibration will be required several times during measurements. Because the sonic source is fixed to the moving objet, problems can occur with reflections of the sonic wave on the moving object structure, in this case on the robot structure. Microphones can detect those reflections and thus give a wrong information on the source position. The reader can find results of the use of ultrasonic systems and extensive comments thereon in Stone[15].

Infrared 3D measurement systems have also been available for a few years. These are based on a stereoscopic analysis by two cameras of the location of an infrared LED fixed to the end effector. Three LED's can be used and switched on and off in sequence so that three fixed points of the end effector can be located. In this way, the absolute position and orientation of a cartesian frame fixed on the end effector can be known. Resolution is about 1/4000 of full scale which depends of relative location of the two cameras. This system needs accurate calibration and is also sensitive to reflections. Such a system was used by Hollerbach[3], and An[4]. The low resolution directly influences the accuracy which, in the judgement of the researchers concerned, was simply not good enough for kinematic calibration. New developments have been carried out (Ishii, Sakane[12]) for infrared 3D measurement systems to bring accuracy up to the level of laser based systems.

Another possible technique of measurement consists in using theodolites and stereo triangulation. This technique was used by Chen[11], Whitney[13] and Judd[14]. Despite the fact that theodolites provide very good accuracy and allow measurements over a wide area of robot workspace, a major drawback is that measurements can only be performed statically and not during the robot motion. It is time consuming and fatiguing, although automatic tracking theodolites exist; moreover, theodolites only measure angles, not lengths, which can cause some problems (need of reference length in particular).

From this review of 3D measurement systems we conclude that it is very difficult to achieve this task mainly for technical reasons but also for economic ones. In fact, it is not necessary to make complete measurement of the end-effector position for calibration purposes. Instead, several equivalent incompletely specified position measurements are sufficient. For example, constraining a point of the end-effector to remain on a given plane or sphere corresponds to one condition. This is realized when a robot tactile sensor is brought into contact with a flat plane or surface of a fixture (cube or spherical surfaces). Another technique consists of constraining the position of lines attached to the end-effector: one possibility is to use an optical sensor in the robot hand and to bring it into a position at which the optical beam just gets broken due to contact with the surface of a fixture (Roth[21]).

For all those reasons we have focused our work on straight line motion analysis.

The measuring device shown in figure 2 was designed taking into account the fact that a straight line motion is easier to track than a general curve and that such a device will be less expensive (see Becquet[8]). The device consists of a magnetostrictive sensor for the longitudinal displacement (along the straight line) and of two magnetic transducers for the two transversal displacements (in a plane perpendicular to the straight line). The range of the longitudinal sensor is 1000 mm and accuracy 0.1 mm. The range of each magnetic transducer is 15 mm and the resolution is 0.6 μm. Considering the complete set-up for measurement, including data acquisition device and measuring device, the accuracy in each direction is about 0.1 mm. The procedure requires minimal human involvement, because it is sufficient to program a straight line motion which follows roughly the axis of the measurement device; automatic data acquisition can be performed easily once the robot carries out with its movement. Complete information for one line can therefore be gathered in a few seconds.

There are two sources of limitations in the use of the measuring device:

- accuracy
- position and orientation of the measuring device with respect to the robot base frame

The second source of limitations is ignored in most papers dealing with this subject.

In fact, it is possible to consider the errors of position and orientation of the measuring device by adding extra parameters in the global model (see section 2.8). It is nothing more than an extra link between the robot base (conventional) and the absolute zero of the measuring device. A total of 6 extra parameters are required for each measuring line:

- 3 parameters for the position error of the measuring device
- 3 parameters for the orientation error of the measuring device.

Of course, these extra parameters will require additional measures to allow their identification.

Another important aspect is that one straight line is not enough to identify the complete set of parameters. For each new line, we will add 6 unknown parameters corresponding to the measuring device position and orientation errors. In section 3.7, some insights for choosing the measuring lines are given.

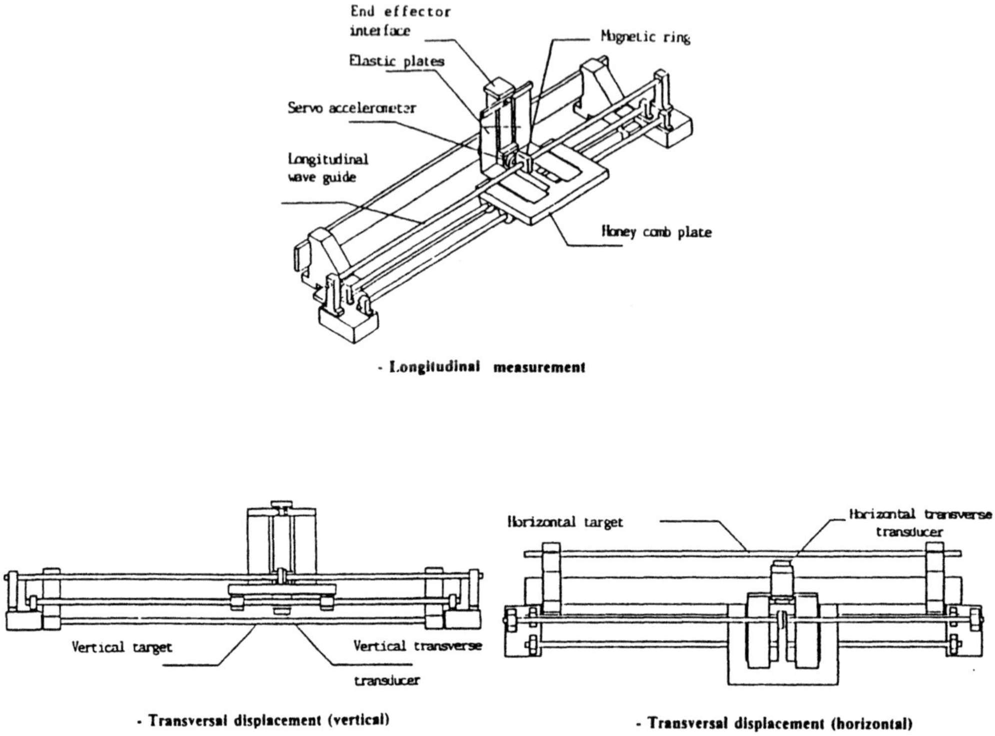

Fig.2. Description of the measuring device: linear motion tracking system with one magnetostrictive sensor (longitudinal displacement) and two magnetic transducers (transversal displacement)

2.4. Differential modeling

The direct geometrical model gives the end-effector position and orientation vector x as a function of the geometrical parameters in a general form given by

$$x = f(\theta, \alpha, a, d, \beta) \qquad \text{(eq. 2.4.1)}$$

where θ, α, a, d and β are R^n-vectors for an n-joint robot.

This relation is strongly nonlinear in the geometrical parameters.

In order to identify them we choose to linearize this model around an initial estimate $\hat{\theta}$, $\hat{\alpha}$, $\hat{a}$, $\hat{d}$, $\hat{\beta}$ (nominal parameters) of the real parameters θ, α, a, d, β. Because two successive parallel axis of rotation are initially supposed to be parallel, we choose $\hat{\beta}_i=0$ for each link.

A differential model can be obtained as follows:

$$\Delta x = J_\theta \Delta\theta + J_\alpha \Delta\alpha + J_a \Delta a + J_d \Delta d + J_\beta \Delta\beta \qquad \text{(eq.2.4.2)}$$

with:

$\Delta\theta$: systematic error vector on the angles θ_i (joint encoder offsets)

$\Delta\alpha$, Δa, Δd, $\Delta\beta$: error vector on the torsion angles α_i ,on the lengths a_i, on the offsets d_i , on the twist angles β_i (all these vectors belong to R^n)

$$J_\theta = \frac{\delta f}{\delta\theta}, \quad J_\alpha = \frac{\delta f}{\delta\alpha}, \quad \textit{etc.}$$

Each of these Jacobian matrices is the sensitivity matrix of the end-effector position and orientation with respect to geometrical parameter variation. These matrices are computed using the initial estimates $\hat{\theta}$, $\hat{\alpha}$, $\hat{a}$, $\hat{d}$, $\hat{\beta}$ for the geometrical parameters. Their computation can be performed by using a symbolic program as presented in Ishii[10] or by evaluating each columns by simple vector relationships (An[4]) such as:

$$\{J_{\theta i}\} = \begin{Bmatrix} [{}^{0}_{i-1}R](\{1z_{i-1}\} x \{{}^{i-1}P_N\}) \\ [{}^{0}_{i-1}R]\{1z_{i-1}\} \end{Bmatrix}$$

where: [], {} and 'x' respectively denote matrix, vector and cross-product

$[{}^{0}_{i-1}R]$ is the rotation matrix from frame (i-1) to the base frame (o)

$\{{}^{i-1}P_N\}$ is the end effector position vector expressed in frame (i-1)

$\{1z_{i-1}\} = \{0, 0, 1\}^t$

and similar expressions for the other Jacobian matrices.

At this stage, computation of the Jacobian matrices is useful for *a priori* analysis: knowing statistical properties of the geometrical parameters (which are related to manufacturing and assembly tolerances), it is possible to establish the statistical properties of the position error of the end-effector, to characterize the distribution of the position errors over the working space to determine working areas with less significant errors and to find the direction of motion in which the smallest position errors will most probably occur (Menq[20]).

Knowing $\hat{\theta}$, $\hat{\alpha}$, $\hat{a}$, $\hat{d}$, $\hat{\beta}$ and computing $\Delta\theta$, $\Delta\alpha$, Δa, Δd, $\Delta\beta$, we can establish a new estimate of the real parameters θ^*, α^*, a^*, d^*, β^* given by:

$$\theta^* = \hat{\theta} + \Delta\theta$$
$$\alpha^* = \hat{\alpha} + \Delta\alpha, \textit{ etc.}$$

Since this is essentially a nonlinear estimation problem, a new geometrical parameter error vector can be recalculated after updating the geometrical parameters, using the same maximum likelihood estimator (see next section), in an iterative way, until the error components fall below some minimum value and the parameters converge to some stable values.

2.5. Maximum Likelihood estimator

The model obtained with equation (2.4.2) that we will write in short

$$\Delta x = J\,\Delta\theta$$

is an ideal differential error model without considering input and output noises.

We now consider the case where only the position error is modelled ($\Delta x \in R^3$).

Here, $\Delta\theta$ represents the complete geometrical parameter error vector, with n components where n is the total number of geometrical parameters to be identified.

Our first assumption (A1) will consist of considering that this model is valid over the complete robot workspace. This implies that we do not take into account non-geometrical errors (backlash, flexibility, *etc.*).

The measures we have to make are:

- the end-effector position error vector Δx_i at the ith point of measure $\in R^3$
- the joint variable vector given by the resolvers $q_i \in R^6$, for a robot with 6 rotational joints.

Both vectors are affected by noise, respectively:

$\delta x_i \quad \in R^3$

$\delta q_i \quad \in R^6$

So, the measured error vector is given by: $\Delta x_i^m = \Delta x_i + \delta x_i$, where the superscript 'm' indicates measured quantities.

Our second assumption (A2) is that δx_i and δq_i are gaussian white noise vectors, with independent components with zero mean and with constant covariance matrices, respectively:

$$S_x = \begin{bmatrix} \sigma_{x1}^2 & 0 & 0 \\ 0 & \sigma_{x2}^2 & 0 \\ 0 & 0 & \sigma_{x3}^2 \end{bmatrix}$$

The values of σ^2_{xi} are obtained from the knowledge of the measuring device accuracy.

$$S_q = \begin{bmatrix} \sigma^2_{q1} & & 0 \\ & \ddots & \\ 0 & & \sigma^2_{q6} \end{bmatrix}$$

The values of σ^2_{qi} are obtained from the knowledge of resolver accuracy and backlash in transmission units (assuming that backlash can be modelled by a gaussian white noise).

Following those assumptions, we can define a linearized noisy model as follows:

$$(\Delta x_i^m - \delta x_i) = J_i \, \Delta\theta + J_{2,i} \, \delta q_i \qquad \text{(eq. 2.5.1)}$$

with $\Delta\theta$: a short form for errors on-geometrical parameters $\varepsilon \ R^n$

J_i: a short form for the global Jacobian matrix $\varepsilon \ R^{3xn}$, function of the nominal parameters

$J_{2,i}$: the manipulator Jacobian $\varepsilon \ R^{3x6}$, function of the nominal parameters

Our third assumption (A3) is that the linearized noisy model is also valid over the complete robot workspace.

The fourth assumption (A4) that we will consider is to assume that the random vector $\Delta\theta \ \varepsilon \ R^n$ follows an *a priori* gaussian distribution, with independent components, with zero mean and with a covariance matrix

$$S_\theta = \begin{bmatrix} \sigma^2_{\theta 1} & & 0 \\ & \ddots & \\ 0 & & \sigma^2_{\theta n} \end{bmatrix}$$

The values of $\sigma^2_{\theta i}$ are obtained from the knowledge of robot link manufacturing tolerances and assembly tolerances.

Finally, we will assume that δx_i, $\Delta\theta$ and δq_i are mutually independent random variables (A5).

Following the Bayesian approach, using the *a priori* distribution of $\Delta\theta$, we can form the likelihood function (Ljung[19]):

$$L = \text{cst.} \exp \{ - \tfrac{1}{2}(\sum_{i=1}^{N} \delta x_i^t \, M_x \, \delta x_i + \sum_{i=1}^{N} \delta q_i^t \, M_q \, \delta q_i + \Delta\theta^t \, M_\theta \, \Delta\theta) \} \qquad \text{(eq. 2.5.2)}$$

with

$$M_x = S_x^{-1},\ M_q = S_q^{-1},\ M_\theta = S_\theta^{-1} \qquad \text{(eq.2.5.3)}$$

and N = number of measurement points

To find the maximum likelihood estimator we have to maximise *L* or to minimise the following cost function:

$$L' = \tfrac{1}{2}\sum_{i=1}^{N} \delta x_i^t M_x \delta x_i + \tfrac{1}{2}\sum_{i=1}^{N} \delta q_i^t M_q \delta q_i + \tfrac{1}{2}\Delta\theta^t M_\theta \Delta\theta \qquad \text{(eq.2.5.4)}$$

To the function L' we have to add the equations of constraints given by the linearized noisy model, using the Lagrange multipliers.

The global cost function over the N measures is given by

$$L = \tfrac{1}{2}\sum_{i=1}^{N} \delta x_i^t M_x \delta x_i + \tfrac{1}{2}\sum_{i=1}^{N} \delta q_i^t M_q \delta q_i$$

$$+ \tfrac{1}{2}\Delta\theta^t M_\theta \Delta\theta + \sum_{i=1}^{N} \lambda_i^t \left[(\Delta x_i^m - \delta x_i) - J_i \Delta\theta - J_{2i}\,\delta q_i\right] \qquad \text{(eq.2.5.5)}$$

The minimization equations lead to:

$$\frac{\partial L}{\partial \Delta\theta} = 0 \implies M_\theta \hat{\Delta\theta} - \sum_{i=1}^{N} J_i^t \lambda_i = 0 \qquad \text{(eq. 2.5.6)}$$

$$\frac{\partial L}{\partial \delta x_i} = 0 \implies M_x \delta x_i - \lambda_i = 0 \qquad \text{(eq. 2.5.7)}$$

$$\frac{\partial L}{\partial \delta q_i} = 0 \implies M_q \delta q_i - J_{2,i}^t \lambda_i = 0 \qquad \text{(eq. 2.5.8)}$$

$$\frac{\partial L}{\partial \lambda_i} = 0 \implies \Delta x_i^m - \delta x_i = J_i \hat{\Delta\theta} + J_{2,i}\delta q_i \qquad \text{(eq. 2.5.9)}$$

Equations (2.5.7) and (2.5.8) yield:

$$M_q \delta q_i = J_{2,i}^t M_x \delta x_i \rightarrow \delta q_i = M_q^{-1} J_{2,i}^t M_x \delta x_i$$

with equation (2.5.9):

$$\Delta x_i^m - J_i \hat{\Delta\theta} = \delta x_i + J_{2,i} M_q^{-1} J_{2,i}^t M_x \delta x_i$$

Thus,

$$\delta x_i = (I + J_{2,i}\, M_q^{-1}\, J_{2,i}^t\, M_x)^{-1}\, (\Delta x_i^m - J_i\, \hat{\Delta\theta})$$

or, in short

$$\delta x_i = A_i(\Delta x_i^m - J_i\hat{\Delta\theta})$$

with

A_i, a 3 x 3 matrix.

Equations (2.5.6) and (2.5.7) yield

$$M_\theta\, \Delta\theta = \sum_{i=1}^{N} J_i^t\, M_x\, \delta x_i = \sum_{i=1}^{N} J_i^t\, M_x\, A_i\, (\Delta x_i^m - J_i\hat{\Delta\theta})$$

which leads to:

$$(M_\theta + \sum_{i=1}^{N} J_i^t\, M_x\, A_i\, J_i)\, \hat{\Delta\theta} = \sum_{i=1}^{N} J_i^t\, M_x\, A_i\, \Delta x_i^m$$

If we call $M_x\, A_i = C_i$, which is a symmetric matrix, equal to:

$$C_i = (M_x^{-1} + J_{2,i}\, M_q^{-1}\, J_{2,i}^t)^{-1},$$

then we have:

$$(M_\theta + \sum_{i=1}^{N} J_i^t\, C_i\, J_i)\, \hat{\Delta\theta} = \sum_{i=1}^{N} J_i^t\, C_i\, \Delta x_i^m$$

and finally:

$$\hat{\Delta\theta} = (M_\theta + \sum_{i=1}^{N} J_i^t\, C_i\, J_i\,)^{-1} \sum_{i=1}^{N} J_i^t\, C_i\, \Delta x_i^m \qquad \text{(eq.2.5.10)}$$

2.6. Recursive computation

As we will see hereafter, it is possible to compute the "least squares solution" corresponding to equation 2.5.10 in a recursive way (Ljung[19]).

Let $\hat{\Delta\theta}(N-1)$ be the estimate of the solution after (N-1) measures.

$$\hat{\Delta\theta}(N-1) = (M_\theta + \sum_{i=1}^{N-1} J_i^{\,t}\, C_i\, J_i)^{-1} \sum_{i=1}^{N-1} J_i^t\, C_i\, \Delta x_i^m \qquad \text{(eq.2.6.1)}$$

By taking

$$P\,(N-1) = (M_\theta + \sum_{i=1}^{N-1} J_i^t\, C_i\, J_i)^{-1} \qquad \text{(eq.2.6.2)}$$

and noting that $P(0) = M_\theta^{-1}$,
we have

$$\hat{\Delta\theta}(N-1) = P(N-1)\ .\ \sum_{i=1}^{N-1} J_i^t\, C_i\ \Delta x_i^m \qquad \text{(eq. 2.6.3)}$$

The estimate of the solution after N measures is now given by:

$$\hat{\Delta\theta}(N) = P(N) \ \sum_{i=1}^{N}\ J_i^t\, C_i\, \Delta x_i^m \qquad \text{(eq.2.6.4)}$$

Equation (2.6.2) leads to:

$$P(N)^{-1} = P(N-1)^{-1} + J_N^t\, C_N\, J_N \qquad \text{(eq. 2.6.5)}$$

Using the matrix inversion lemma (Ljung[19]):
$[A+BCD]^{-1}=A^{-1}-A^{-1}B[DA^{-1}B+C^{-1}]^{-1}DA^{-1}$,

it can be shown that

$$P(N) = P(N-1) - P(N-1)\ .\ J_N^t\,(C_N^{-1} + J_N\, P(N-1)\, J_N^t)^{-1}\ .\ J_N\, P(N-1) \qquad \text{(eq.2.6.6)}$$

Equation (2.6.6) is used to compute recursively the matrix P(N).

Equations (2.6.1) and (2.6.5) yield:

$$\sum_{i=1}^{N-1} J_i^t\, C_i\, \Delta x_i^m = P(N-1)^{-1}\ \hat{\Delta\theta}(N-1) = P(N)^{-1}\hat{\Delta\theta}(N-1) - J_N^t\, C_N\, J_N\, \hat{\Delta\theta}(N-1) \qquad \text{(eq.2.6.7)}$$

Equations (2.6.4) and (2.6.7) give:

$$\hat{\Delta\theta}(N) = P(N)\ .\ (\sum_{i=1}^{N-1} J_i^t\, C_i\, \Delta x_i^m + J_N^t\, C_N\, \Delta x_N^m)$$

$$= P(N)\ (P(N)^{-1}\ \hat{\Delta\theta}(N-1) - J_N^t\ C_N\ J_N\ \hat{\Delta\theta}(N-1) + J_N^t\ C_N\ \Delta x_N^m)$$

We thus arrive at the algorithm:

$$\Delta\theta(N) = \Delta\theta(N-1) + P(N)\ J_N^t\ C_N[\Delta x_N^m - J_N\ \hat{\Delta\theta}(N-1)] \qquad \text{(eq. 2.6.8)}$$

Equation (2.6.8) can be used to compute recursively the estimate of the solution as a function of the previous estimate $\hat{\Delta\theta}(N-1)$, the matrix P(N) and the prediction error $(\Delta x_N^m - J_N\hat{\Delta\theta}(N-1))$

2.7. Properties of the recursive estimator

Multiplying equation (2.5.1) by $J_i^t C_i$ and summing over all measures, we can obtain the following result:

$$(M_\theta + \sum_{i=1}^{N} J^t{}_i\ C_i\ J_i)\ \Delta\theta = \sum_{i=1}^{N} J_i^t\ C_i\ \Delta x_i^m - \sum_{i=1}^{N} J_i^t\ C_i\ \delta x_i - \sum_{i=1}^{N} J_i^t\ C_i\ J_{2,i}\ \delta q_i + M_\theta \Delta\theta \qquad \text{(eq. 2.7.1)}$$

We will now define

$$\tilde{\theta}_N = \hat{\Delta\theta}(N) - \Delta\theta \qquad \text{(eq.2.7.2.a)}$$

the *a posteriori* estimation error, and

$$\tilde{\theta}_o = \hat{\Delta\theta}(0) - \Delta\theta = 0 - \Delta\theta \qquad \text{(eq.2.7.2.b)}$$

the *a priori* estimation error.

Subtracting equation (2.7.1) from equation (2.5.10) we obtain:

$$(M_\theta + \sum_{i=1}^{N} J_i^t\ C_i\ J_i)\ \tilde{\theta}_N = \sum_{i=1}^{N} J_i^t\ C_i\ \delta x_i + \sum_{i=1}^{N} J_i^t\ C_i\ J_{2,i}\ \delta q_i + M_\theta \tilde{\theta}_o \qquad \text{(eq. 2.7.3)}$$

Equation (2.7.3) leads to:

$$\tilde{\theta}_N = P(N) \,.\, [\sum_{i=1}^{N} J_i^t C_i \delta x_i + \sum_{i=1}^{N} J_i^t C_i J_{2,1} \delta q_i + M_\theta \tilde{\theta}_o] \qquad \text{(eq. 2.7.4)}$$

Following our assumptions (A2), (A4) and (A5),

$$<\delta x_i> = 0$$

$$<\delta q_i> = 0$$

$$<\tilde{\theta}_o> = 0$$

(where the symbol '< >' means mathematical expectation)

we have $<\tilde{\theta}_N> = 0$, which demonstrates that the estimator is unbiased.

We can also show that the variance of the *a posteriori* estimation error is equal to P(N):

$$<\tilde{\theta}_N \tilde{\theta}_N^t> = P(N) [\Sigma J_i^t C_i M_x^{-1} C_i^t J_i +$$

$$\Sigma J_i^t C_i J_{2,i} M_q^{-1} J_{2,i}^t C_i^t J_i + M_\theta M_\theta^{-1} M_\theta] P(N) \qquad \text{(eq. 2.7.5)}$$

$$<\tilde{\theta}_N \tilde{\theta}_N^t> = P(N) [\Sigma J_i^t C_i(M_x^{-1} + J_{2,i} M_q^{-1} J_{2,i}^t) C_i^t J_i + M_\theta] P(N)$$

$$= P(N) [\Sigma J_i^t C_i I J_i + M_\theta] P(N)$$

$$= P(N) P(N)^{-1} P(N) = P(N). \qquad \text{(eq. 2.7.6)}$$

2.8. Global model

The global model is obtained by adding to the model represented by equation (2.5.5) the effect of position and orientation errors with respect to the robot base frame of the measuring device.
The contribution to the error on the robot end-effector position for the ith measure due to a position error Δx^d_j, Δy^d_j, Δz^d_j and orientation error $\Delta\phi^d_{x\,j}$, $\Delta\phi^d_{y\,j}$, $\Delta\phi^d_{z\,j}$ of the jth measuring line is obtained by:

$$e_i = J_{i,j}^d \Delta x^d_j \qquad \text{(eq. 2.8.1)}$$

with Δx^d_j: a short form for Δx^d_j, Δy^d_j, Δz^d_j, $\Delta\phi^d_{x\,j}$, $\Delta\phi^d_{y\,j}$, $\Delta\phi^d_{z\,j}$ (ε R^6)

(superscript 'd' indicates values related to the measuring device)

$J^d_{i,j}$: the measuring device Jacobian matrix for the ith measure on the jth measuring line (ε R^{3x6})

The matrix $J^d_{i,j}$ is given by:

$$J^d_{i,j} = (I_3, -\hat{P}_{j,i}) \qquad \text{(eq. 2.8.2)}$$

with I_3: the unit matrix (ε R^{3x3})

$\hat{P}_{j,i}$: the skew symmetric matrix (ε R^{3x3}) corresponding to the cross-product $P_{j,i} \times \Delta\phi$, with $P_{j,i}$ the coordinates of the ith measurement point on the jth line with respect to the measuring device reference frame: $\hat{P}_{j,i}\ \Delta\phi \triangleq P_{j,i} \times \Delta\phi$

Thus, equation (2.5.1) becomes, for the global model:

$$(\Delta x_i^m - \delta x_i) = J_i\, \Delta\theta + J_{2,i}\, \delta q_i + J_{i,j}{}^d\, \Delta x_j{}^d$$

if the point i is located on the jth measuring line.

We can, consequently, choose an extended parameter vector $\Delta\theta$, including $\Delta x^d{}_j$ for all j, and an extended Jacobian matrix J_i, including $J_{i,j}{}^d$ (which is non-zero only if the ith measurement point belongs to the jth measuring line).
All our results and developments are therefore also valid for the global model.

2.9. Comparison with other identification techniques

Our maximum likelihood estimator is a generalized least squares identification algorithm, including a probabilistic characterization of measurement noise and of unknown robot parameters (*a priori* distributions). The common least-squares algorithm, which has always been adopted in previous calibration works (see Veitschegger[1], An[4], Preumont[5] and Judd[14]), is a particular case of the maximum likelihood estimator with:

$S_x = I_3$ (3x3 identity matrix)
$S_q = 0$ (joint encoders readings are assumed error free)
$S_\theta = P(0) = P_0 I_n$ (P_0 is an arbitrary large number)

In equations (2.6.6.) and (2.6.8.), the matrix C_N is then the identity matrix and these equations become identical to the recursive least-squares formulas (Ljung[19], p. 20).
From this point of view, we can conclude that our method is a better approach, because in a Bayesian framework, it takes into account:

- *a priori* information on the output observations ($S_x \neq I_3$)
- *a priori* information on the input readings ($S_q \neq 0$)
- *a priori* information on the parameter vector (S_θ), which is related to the initial condition of the covariance matrix P(0) (Bayesian interpretation).

It is well known that using all available prior information gives better results in convergence and stability (Ljung[19], pp.299-303). For example, if initial values θ_0 are known with good precision, large value of P(0) would cause abrupt transients and a more difficult convergence. On the other hand, when P(0) is too small (too much confidence given to the erroneous initial values), the convergence can be very slow. This suggests that, for a given finite number of measured data, the identification results will be more accurate when *a priori* information is employed. These considerations will be illustrated by an example on a real case in section 3.

3. Application

3.1. Study case

The kinematic calibration algorithm was applied to one of our robots, developed in our laboratory, which has degrees of freedom (see fig.3). This robot is controlled by a SIEMENS RCM-SIROTEC control unit. In this system, most of the geometrical parameters are factory programmed and can not be modified by the user. The consequences of this fact will be investigated subsequently.

3.2. Nominal values

Table 1 gives the nominal values of the geometrical parameters.

Table 1 - Nominal Values of Geometrical Parameters (angles in rad., lengths in m.)

i	α_i	a_i	d_i	θ_i
1	$-\pi/2$	0	0	θ_1
2	0	0.350	0	θ_2
3	$\pi/2$	0	0	θ_3
4	$-\pi/2$	0	0.350	θ_4
5	$\pi/2$	0	0	θ_5
6	0	0	0.1	θ_6

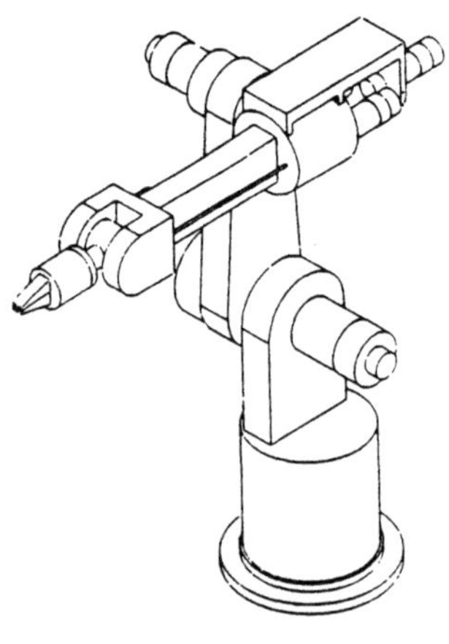

Fig.3. ULB-Mainpulator (6 degrees of freedom)

3.3. Covariances matrices

The following values based on measuring device accuracy, resolver resolution and manufacturing tolerances were used in the different covariance matrices:

$$S_x = 10^{-9}\begin{bmatrix} 3.84 & 0 & 0 \\ 0 & 10 & 0 \\ 0 & 0 & 3.84 \end{bmatrix}[m^2]$$

$$S_q = 10^{-9}\begin{bmatrix} 21 & 0 & 0 & 0 & 0 \\ 0 & 21 & 0 & 0 & 0 \\ 0 & 0 & 21 & 0 & 0 \\ 0 & 0 & 0 & 1.31 & 0 \\ 0 & 0 & 0 & 0 & 305 \end{bmatrix}[rad^2]$$

(the high value of the last element is due to backlash in the last joint)

For S_θ, on the different lengths (a_i, d_i, $\Delta x^d{}_j$, $\Delta y^d{}_j$, $\Delta z^d{}_j$)

$\sigma = 0.001$ m

and on the different angles (α_i, β_i, $\Delta\phi^d_{x\,j}$, $\Delta\phi^d_{y\,j}$, $\Delta\phi^d_{z\,j}$)

$\sigma = 3° = 0.052$ rad

3.4. Remarks

Before presenting our results, we have to make some remarks and comments.

Non uniqueness of the solution:

If we consider the global set of geometrical parameters one can point out that the orthogonality of different parameters is far from being guaranteed. Indeed, some parameters are not even independent. For example, an error on θ_1 has the same effect on the end-effector position as an error on the measuring device position and orientation. In a similar way, an error on d_2 or on d_3 has the same effect on the end- effector.

Non uniqueness of the solution is not a real problem considering the user's point of view. The obtained solution will guarantee a certain level of accuracy according to a least square fit of the parameters. But, surprisingly, if the global geometrical model is satisfactory, the errors identified, taken one by one, do not have a real physical meaning.

Furthermore, it can be shown that, in the Jacobian matrix, the columns associated with the geometrical parameters are far from being orthogonal to the ones associated with the non-geometrical errors. Therefore, if non-geometrical errors are not included in the estimation, their effects will be distributed over geometrical parameters, so that the attribution of estimated values to specific physical errors may be meaningless.

Fixed parameters
Following the same idea, it is important to try to find some parameters that could not be identified. This can be done by carefully examining the robot kinematic structure. In our case we have decided to "freeze" the following geometrical parameters:

θ_1: an error on θ_1 has the same effect as some position and orientation error of the measuring device
α_6: an error on α_6 is not observable
d_1: an error on d_1 has the same effect as Δz^{d}_{j}, device position error in the vertical direction
d_2: an error on d_2 has the same effect as an error on d_3.

On the other hand, a twist error β_2 must be introduced, to take into account a non-exact parallelism between axes z_1 and z_2.

Both remarks are also presented in Veitschegger [1], Becquet[7] and Chen[11].

3.5. Errors on electrical zeroes of joint encoders

Our first results are related to a case in which we only consider errors on the joint encoder zeroes: we consider $\Delta\theta_i = 0$ and $\Delta a_i = \Delta d_i = \Delta\alpha_i = \Delta\beta_i = 0$ for all i.

As we know from Whitney[13] and Judd[14], the joint angle offsets are by far the largest contributor to the end effector error (about 90%), so that a significant improvement of the geometrical model can be expected after identification.

During the experimental work, we measured 12 points on each line, with a total of 5 lines in the workspace. One of those lines is only used for validation of the identification results.

Before identification, the mean end-effector position error on the total set of measurements belonging to the first 4 lines is equal to:

$$\bar{e} = 11.28 \text{ mm}$$

After a first run of the identification algorithm, the average error is equal to:

$$\bar{e}_1 = 4.95 \text{ mm}$$

To improve our results it is appropriate to use our first order linearized model in a next iteration, as we said in section 2.4, taking the covariance matrix P_N of the previous run for P_o of the next run, and the estimates θ_N of the previous run for the initial guess θ_o of the next run.
After the second run the average error is equal to:

$$\bar{e}_2 = 0.89 \text{ mm}$$

After the third run the average error is equal to:

$$\bar{e}_3 = 0.81 \text{ mm}$$

Further runs do not improve the solution.

Errors identified are as follows (in rad.):

$\Delta\theta_2 = 0.0236$
$\Delta\theta_3 = -0.1277$
$\Delta\theta_4 = -0.0341$
$\Delta\theta_5 = 0.2585$
$\Delta\theta_6 = -0.0109$

The square roots of the diagonal elements of the final covariance matrix are respectively:
$\sigma_2 = 0.0005$ rad
$\sigma_3 = 0.0007$ rad
$\sigma_4 = 0.0014$ rad
$\sigma_5 = 0.0010$ rad
$\sigma_6 = 0.0148$ rad

On the validation line, we have a mean end-effector position error equal to
- before identification:
  $\bar{e}_V = 10.83$ mm
- using the results of identification:
  $\bar{e}_{V3} = 0.81$ mm

which is very close to the results obtained for the identification lines.

3.6. User adjustable geometrical parameters on SIEMENS control unit

As already mentioned, the SIEMENS RMC-SIROTEC control unit that we use with our home made robot allows the user to adjust only a subset of the geometrical parameters. Therefore, we search for a set of values for the adjustable parameters that best fit the data included in the estimation process; we must keep in mind the fact that this set of geometrical values may be valid only in a limited area of the robot workspace.

In our case, we can adjust
the angular offsets for $\theta_1, \theta_2, \ldots, \theta_6$,
and the link lengths a_2, a_6, d_4, d_6

The parameter θ_1 will not be adjusted (see remarks in section 3.4).

We have used the same measurements as in section 3.5.
After the first run of the algorithm, the mean error is equal to:

$$\bar{e}_1 = 3.15 \text{ mm}$$

After the 12th run, the mean error is equal to:

$$\bar{e}_{12} = 0.61 \text{ mm}$$

Further runs do not improve the solution.
Identification results are:

$\Delta\theta_2 = 0.0049$ rad	(0.0002 rad)
$\Delta\theta_3 = -0.0995$ rad	(0.0005 rad)
$\Delta\theta_4 = -0.0334$ rad	(0.0010 rad)
$\Delta\theta_5 = 0.1760$ rad	(0.0012 rad)
$\Delta\theta_6 = 0.3122$ rad	(0.0122 rad)

$\Delta a_2 = -0.0022$ m	(0.00011 m)
$\Delta d_4 = -0.0148$ m	(0.00034 m)
$\Delta a_6 = 0.0017$ m	(0.00008 m)
$\Delta d_6 = -0.0166$ m	(0.00010 m)

Values in brackets are the square-roots of the covariance matrix diagonal elements.
On the validation line, we have, using the results of the identification, a mean end-effector position error equal to:

$$\overline{e}_{v12} = 0.68 \text{ mm},$$

which is of the same order as the mean error for the four identification lines.

3.7. Comments

1) The improvement achieved by taking into account Δa_2, Δd_4, Δa_6 and Δd_3 is about 25%.

2) The difference on some values of error between the two cases is significant.

	1st case (rad)	2nd case (rad)
$\Delta\theta_2$	0.0236	0.0049
$\Delta\theta_5$	0.2585	0.1760
$\Delta\theta_6$	-0.0109	0.3122

This fact clearly illustrates the difficulty of a physical interpretation of the identification results, because of the non-completeness of the model adopted.

3) Experimental results clearly show that a lower bound exists on the calibration error that is dictated by robot repeatability and calibration accuracy. The latter depends in turn on the accuracy of the measurement system and the inaccuracies due to robot parameters that are not modelled (non-geometrical errors).

4) The choice of the measuring lines is dictated by two considerations. The first one is that there exist some areas in the robot workspace where the position error is large. For calibration purpose, it is interesting to explore such regions which can be found by an *a priori* analysis (see section 2.4). The reason is that measurement errors become less significant. The second consideration is that the set of the geometrical values identified can be valid in a limited area of the robot workspace, because an incomplete set of parameters has been adopted in the modeling process and no non-geometrical effects have been taken into account.

This suggests that the measuring lines must be chosen in a restricted region of the workspace, particularly in the zones where the task must be performed with great accuracy.

5) We have compared the performances of the recursive maximum-likelihood (M-L) estimator, against the recursive least-squares (L-S) algorithm, by running the latter on the same measurement data set. The initial estimate of the covariance matrix P(0) (=$P_0 I_n$) was chosen such that convergence rate was optimal. Figure 4 shows the module of the prediction error (see section 2.6. for the definition) during the identification process.

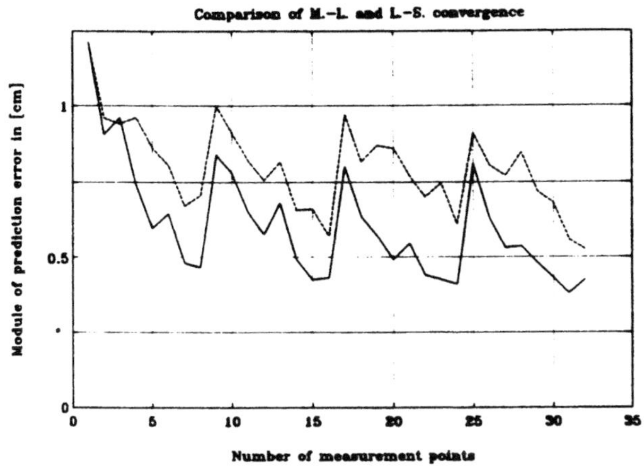

Fig.4. Comparison of Maximum-Likelihood and Least-Squares algorithm convergences

Notice that:

a) each measuring line involves ten points. The first two points are used to compute an initial estimate of the position and the orientation of the line, so that 8 points are effectively used for identification;

b) each time that a new measuring line is taken up, the prediction error abruptly increases, because several extra parameters (related to position and orientation error of the line) must be identified;

c) prediction errors obtained by the M-L algorithm are smaller and, after four lines have been explored, the accuracy of the estimation is better than with the L-S algorithm. However, we observed that the accuracy levels become nearly identical if a large number of measurement points is used.

4. Use of Improved Geometrical Model

The improved geometrical model obtained after identification of the geometrical parameter errors can be used in different ways.

4.1. Error map

The error map is obtained by plotting, in a cross-section of the robot workspace, the end-effector position error vector norm, based on the knowledge of the nominal geometrical model and the improved model.

We call the nominal position vector of the end-effector x_n and the position vector of the end-effector computed by using the identified geometrical parameters x_i. Both terms are defined with respect to the robot frame. These quantities are functions of q_i:

$$x_n = f_n(q) \qquad \text{(eq. 4.1.1)}$$

$$x_i = f_i(q) \qquad \text{(eq. 4.1.2)}$$

with f_n: nominal geometrical model
f_i: improved geometrical model

The error vector norm is obtained by:

$$e = \| x_i - x_n \| = \|f_i(q) - f_n(q)\| \qquad \text{(eq.4.1.3)}$$

Plotting e as a function of the cartesian coordinates in a workspace cross section gives us information on subspaces, showing which are more or less accurate than other ones (see figure 5 representing an error map in a vertical cross section of our robot workspace).

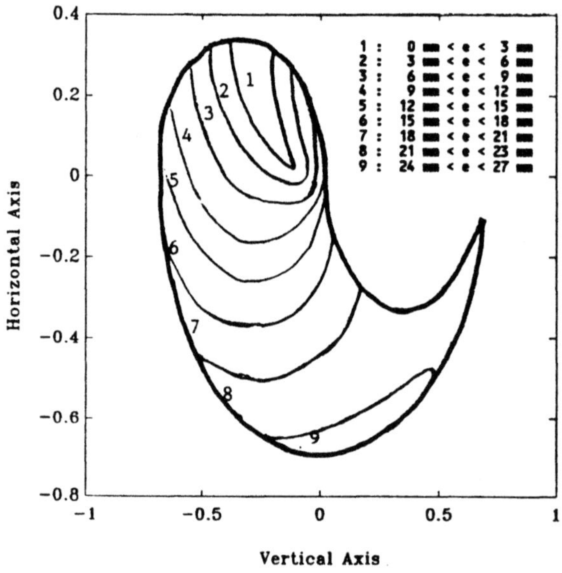

Fig.5. Error map in a vertical cross section of the ULB-manipulator workspace (θ_1=θ_4=θ_6=0 - elbow up)

This information is very helpful for the design of the layout of a robot cell when the robot control system and the off-line programming system do not permit compensation for the end-effector errors by using the other methods here described. The idea of this error map is a continuation of the concept of *a priori* analysis and *a posteriori* analysis of geometrical model errors presented in Becquet[7].

4.2. Action on the target: task space correction

If we cannot modify the geometrical model inside the robot control unit, but we can adapt the off-line programming system, then we may change the target in the task space in such a way that the end-effector will reach the initial desired target x_d (Payannet[9]): this is the "false target" technique.

Let us call the desired target: x_d
the target reached by the end-effector without correction ($\neq x_d$): x_n
the modified target: x_m

With a very good approximation, x_m is given by:

$$x_m = x_d + (x_d - x_n) = 2\,x_d - x_n = 2\,x_d - f_i(f^{-1}{}_n\,(x_d)) \qquad \text{(eq. 4.2.1)}$$

The modified target x_m is presented to the control unit which computes:

$$q_m = f_n^{-1}\,(x_m) \qquad \text{(eq. 4.2.2)}$$

This type of algorithm requires to handle 2 inverse nominal models and one direct improved model. Similar algorithms are presented in Veitschegger[1] and Preumont[5], while Judd[14] proposed an improved iterative version of this one.

4.3. Correction in the joint space

Obviously, the best way to use the improved geometrical model is to replace the inverse nominal geometrical model by the improved one in the robot control unit. But, by doing so, it is no longer possible to find an analytical solution for the inverse geometrical problem. Therefore, iterative methods (like Newton- Raphson methods) must be implemented in the control unit, in order to convert the cartesian task into an articular one, using the improved geometrical model (Cai[2]).

If the inverse geometrical model of the control unit cannot be improved directly, but we can control the robot directly in the joint variable space, then we can use the following algorithm to compensate for the error (Payannet[9]):

$$q_i = f^{-1}{}_n\,(x_d) + J_n^{-1}\,[\,x_d - x_n] = f^{-1}{}_n\,(x_d) + J_n^{-1}\,[\,x_d - f_i(f^{-1}{}_n\,(x_d)] \qquad \text{(eq. 4.3.1)}$$

where J is the classical Jacobian matrix of the manipulator and q_i, the input of the robot controller. This algorithm requires one inverse nominal model, one improved direct model and one inverse Jacobian of the nominal model.

If the identification results show that the errors on the electrical zeroes of the encoders are responsible for the greatest part of the total end-effector error, and if the robot can be controlled in the joint space, we propose simply:

$$q_i = f^{-1}{}_n\,(x_d) - \Delta q \qquad \text{(eq. 4.3.2)}$$

where q_i: inputs of the robot controller
Δq: identified angular offset errors.

This technique only requires us to compute the inverse of the nominal model. Moreover, this computation can be performed off-line.

5. Conclusions of Part I

The proposed method, including the algorithm and the measuring device, shows very good results in improving the accuracy of the robot model. In our opinion, an error of 0.5 mm must be considered as the best result that can be obtained without considering the non-geometrical errors. During our experiments, we have loaded the end-effector with a load of one kilogram: the resulting deflection was 0.2 mm. Furthermore, backlash introduces errors of the same order. These facts explain our figure of 0.5 mm.

The algorithm does not consider only the errors that have to be identified but also the input and output noises. It presents a recursive form with a reasonable amount of computation. It offers the possibility of following the convergence rate and direction of the solution step by step, and of detecting and eliminating possible faulty measurements.

We have summarized the different ways to use the improved geometrical model.

Now, efforts should be focused on the non-geometrical errors, with the aim of bringing the accuracy of the system close to the resolution of the encoders. This is the purpose of Part II.

PART II: NON-GEOMETRICAL PARAMETERS IDENTIFICATION AND ERROR COMPENSATION BY NEURAL NETWORK ESTIMATOR

1. Introduction

Modeling and identification of non-geometrical errors are not so common in the field of robot calibration. There are only a few works dealing with this problem: Withney[13] uses a model including non-geometrical errors and a least squares numerical search algorithm on theodolite measurements, but the error model is completely non-linear; moreover, the definition of parameters to be identified is rather unusual and related to measuring instruments. Judd[14], on the other hand, considers mainly non-geometrical errors (gear train errors, joint and link flexibility, etc.) and proposes error models which can be used for identification with a common least squares procedure; unfortunately, the quality of the identification results strongly depends on the presumed structure of the model.

In this second part of the paper, we will assume that the nominal model is already tuned to compensate for the geometrical error effects (see Part I); then, the real position reached by the end-effector is obtained by adding to the nominal model output the contribution of a residual error function containing all the non-geometrical effects. As explained in Renders[25], we use a neural network (NN) in order to identify this error function, without the need of an explicit modeling stage, while taking advantage of the potential computational benefits expected from this new approach (parallel computation, adaptation to large number of parameters, natural fault tolerance, natural robustness, generalization capacity, etc.). A supervised learning process, based on the back-propagation algorithm, is applied to develop the network.

The following section of the second part of the paper briefly describes the main sources of positioning error in robot kinematics and introduces the residual error function. Section 3 develop the neural network architecture and the training stage methodology. Then, different ways to use the neural network's results for correction and compensation are presented in section 4. Results of simulations on a robot with 3 degrees of freedom illustrating the performance of the network for calibration purposes are shown in section 5. Finally, section 6 contains some conclusions about the work.

2. Position Error Sources

2.1. Geometrical errors

A classical choice for the geometrical parameters used to describe the robot geometry is the Denavit and Hartenberg parameters; for kinematic identification and calibration, it is usually necessary to introduce an extra parameter (the link twist, bi) in the case of successive nominally parallel rotational joints (Veitschegger[1]).

Nominal values of those geometrical parameters can be different from the real ones in the robot system, because manufacturing and assembly tolerances affect the kinematic model of the robot; moreover, the electrical zeroes of the joint encoders do not generally coincide with the mechanical zeroes of the joints themselves. Usually, geometrical errors on geometrical parameters are responsible for about 90% of the total error, while the remaining 10% are due to non-geometrical errors. As previously mentioned, in the following we consider that the nominal geometric model is already tuned, so that only non-geometrical errors will be identified and compensated for.

2.2. Non-geometrical errors

As already mentioned in the first part, the sources of non-geometrical errors lie with:
-joint flexibility
-link flexibility
-structural deformation of the base
-gear transmission error (runout and orientation error)
-backlash in gear transmission
-temperature effect.

Link flexibility is usually less than joint flexibility (below 5%); the flexibility of the transmission unit can be considered as a torsional spring (or linear spring for prismatic joints) inserted between the output of the gear box (assumed as rigid) and the link drived by the transmission.

Gear runout error occurs when the actual center of rotation does not lie at the center of the pitch circle. It produces a cyclic error depending on the angle of rotation of the gear and the reduction ratio. Orientation error is a second important error existing in gear trains; it appears when the two axis of the gears are not exactly parallel: the contact line of one gear with respect to the other is no longer circular, but ellipsoidal; this error is cyclic with a period that is half a revolution. Details on this kind of errors are explained in Judd[14].

Backlash is probably one of the most difficult error source to identify. We will assume that the backlash is closed; it is true if the torque applied by the actuator is larger than the friction torque; then, following the direction of the applied torque, the gear will be at one or the other extreme of the dead zone, even if this torque is due to gravity in static condition. Otherwise, backlash should be considered as a noise (with roughly uniform distribution) perturbating the accuracy of the mechanism.

The effect of temperature is to expand the robot mechanical structure. Given the materials used for robot links and their thermal expansion coefficient, the error due to temperature in traditional industrial environment is responsible for 0.1% of the total error. However, in several situations of teleoperation (nuclear plants, spatial applications, ...) the effect of the temperature can be more considerable.

2.3. Residual error function

Let us call θ the vector of the joint encoder angles; these angles are measured directly on the rotor shaft of each motor, prior to backlash, gear runout and flexibility effects. Let us call f_n, the nominal geometrical model(already tuned to take into account and to compensate for the effects of the geometrical errors, following techniques of Part I), that gives the nominal vector of the end effector position/orientation x_n as a function of θ:

$$x_n=f_n(\theta) \tag{1}$$

Let us call x_r the vector of the real position/orientation of the end effector, which differs from x_n because of non-geometrical errors, and f_r the function of x_n and θ which produces x_r:

$$x_r=f_r(x_n,\theta)=f_r(f_n(\theta),\theta) \tag{2}$$

Introducing the residual error function $\varepsilon(\theta)$ defined by:

$$\varepsilon(\theta)=f_r(f_n(\theta),\theta)-f_n(\theta) \tag{3}$$

we have the following relationship:

$$x_r=x_n+\varepsilon(\theta) \tag{4}$$

The objective consists then of trying to identify (or to approximate) this residual error function $\varepsilon(\theta)$ having a set of measurement data (θ,x_r). Note that this function also depends on the payload carried by the robot, because of the gravity torque and the flexibility. In our work, we have chosen to perform this identification task by training a neural network with supervised learning.

3. Neural Network: Architecture and Training

An artificial neural network consists of many simple processing elements called units that interact using weighted connections. Each unit has a state or activity level that is determined by the input received from other units in the network.

The short-term knowledge of a neural network is usually encoded by the states of the units. All the long-term knowledge is codified by the weights. Thus, learning amounts to modify the weights in a way that allows the networks to construct the necessary internal representations of their environments in order to solve the task they face.

Recently, it has been proved that multilayer feedforward networks with at least one hidden layer of sigmoid units (see below) and using a learning algorithm of the backpropagation type [22] can approximate any square-integrable function to any desired degree of accuracy, provided sufficient hidden units are available [23]. Nevertheless, there are no guarantees that the solution can be found. In addition, the optimal number of weights is not known in advance, what could result in a network larger than required which, consequently, arbitrarily approximates the training data but generalizes poorly. The learning algorithm we have used addresses simultaneously these two topics.

In a multilayer feedforward network, units are arranged into layers in such a way that a given unit can only send its output to units located in layers above. In this manner, a network is made of an input layer (bottom), one or more hidden layers (middle) and an output layer (top). The learning phase is aimed at reducing the difference between the actual output of the network and the desired one.

An input unit simply forwards the information it receives from outside. Two are the kinds of input signals to our network: either a joint angle, a real number in the interval $[-\pi,\pi]$, or a payload, a positive real number.

A hidden unit computes its output by applying a sigmoid function to its activation level:

$$o_i = \frac{2}{1+e^{-\beta \left(\sum_j w_{ij} o_j + \sigma_i\right)}} - 1$$

where o_i is the output of the unit i, σ_i is a bias, j ranges over all the units sending their outputs to the unit i, w_{ij} is the weight associated to the connection from the unit j to the unit i, and β is a positive constant (0.4 in the simulations reported here). We have chosen a symmetrical sigmoid in order to avoid rapid saturation with extreme inputs that makes learning even more difficult.

An output unit produces as output its activation level.

The learning algorithm we have used is characterized by two processes. The first of them is the rule for updating the weights. The weights are modified according to a variant of the basic backpropagation algorithm:

$$\Delta w_{ij} = -\eta_i \frac{\delta E}{\delta w_{ij}}$$

where η_i is the learning rate associated to the unit i and E is the error function to be minimized. Three are the modifications we have introduced. Firstly, the weights are not updated after the gradients have been accumulated over the whole training set, but after the presentation of each pattern. This version, known as stochastic or on-line version, allows to exploit redundancy in the training data (as it is our case: see Section 5, where it is described how the training data were obtained).

Secondly, the learning rate is specific for each unit:

$$\eta_i = \frac{\alpha}{\rho_i}$$

where α is a constant and ρ_i is the number of input connections to the unit i [24].

Thirdly, the value of α decreases as learning proceeds until reaching a sufficiently small value. At this moment the learning process has converged (successfully or not). Nevertheless, we have observed that if, at this moment, the value of α is risen and the process of incrementally reducing α is repeated, then the weight configuration is better than before. A possible explanation is that when the learning rate is increased, the weight configuration is perturbed sufficiently as to climb the walls of the basin of attraction it was in. Experiments confirm this explanation since the error increases when the learning rate increases. In the simulations, we have iterated the cycle "set α to its initial value (0.2) --- decrease gradually α up to a final value (0.001)" until the network does not improve significatively.

The second feature of our learning algorithm is that the network is built incrementally. A new hidden unit is added (either in an existen hidden layer or in a new one) when the current network cannot reduce any more the error. By modifying the network architecture, the shape of the weight space is also changed, what could remove the local minimum the network was trapped in. Experiments confirm this fact: after adding a new hidden unit, the performance of the network always improves. In addition, this technique could allow to build the network with the smallest number of weights required to solve the task at hand. Consequently, it is expected to obtain a network with good generalization abilities. Again, the experiments we have carried out confirm this hypothesis.

Finally, it is worth noting that we have observed that the performance of a network built incrementally is always better than the performance of an equivalent network trained from the beginning with the final architecture.

4. Positioning Error Compensation and Trajectory Correction

In this section, we will present two ways to compensate for the effects of non-geometrical errors, using the output $\varepsilon(\theta)$ of the neural network, which is nothing else that the estimate of the residual error function $\varepsilon(\theta)$.

4.1. Action on the target: task space correction

The idea is to define a false target in such a way that the end effector will reach a position closer to the desired one (Payannet[9]).
Let us call:

x_d, the desired position/orientation

x_r, the real position/orientation of the robot

x_m, the modified position/orientation (false target)

f_n, the nominal direct geometrical model

f_n^{-1}, the nominal inverse geometrical model (implemented in the robot control unit).

If we present to control unit of the robot the modified target given by:

$$x_m = x_d - \hat{\varepsilon}(f_n^{-1}(x_d)) \tag{5}$$

the robot will reach the following position/orientation:

$$x_r = x_m + \varepsilon(f_n^{-1}(x_m)) \qquad \text{(using eq. 5)}$$

$$= x_d + \varepsilon(f_n^{-1}(x_m)) - \hat{\varepsilon}(f_n^{-1}(x_d)) \tag{6}$$

$$\cong x_d \quad \text{since, with a very good approximation, } \varepsilon(f_n^{-1}(x_m)) \cong \hat{\varepsilon}(f_n^{-1}(x_d)) .$$

Then, the control unit computes:

$$\theta_m = f_n^{-1}(x_m) \tag{7}$$

This type of algorithm requires 2 nominal inverse model calls and one neural network computation.

4.2. Correction in the joint space

If the control unit allows to control the robot directly in the joint space, then we can use the following algorithm to compensate for the positioning error (which is nothing else that the differential version of the preceding one):

$$\theta_m = f_n^{-1}(x_d) - J_n^{-1} . \hat{\varepsilon}(f_n^{-1}(x_d)) \tag{8}$$

where J_n is the Jacobian matrix of the robot, computed in $\theta_d = f_n^{-1}(x_d)$.

Indeed, using eq. 8, the robot will reach the position/orientation given by:

$$x_r = f_n(\theta_m) + \varepsilon(\theta_m) \qquad \text{(using eq. 5)}$$

$$= x_d - J_n \cdot J_n^{-1} . \hat{\varepsilon}(f_n^{-1}(x_d)) + \varepsilon(\theta_m) \quad \text{(developing } f_n \text{ up to the first order)} \tag{9}$$

$$\cong x_d \quad \text{since, with a very good approximation, } \varepsilon(\theta_m) \cong \hat{\varepsilon}(f_n^{-1}(x_d)) .$$

Note that both algorithms use the same approximation, but the second one uses one nominal inverse model, one inverse Jacobian (which could be sometimes easier to compute than the inverse model) and one neural network computation.

5. Case Study - Results of Simulation

Simulations were performed on a robot with 6 degrees of freedom; the end effector is defined as the intersection point of the three last axis (see Fig.3). Only the end effector position (x,y,z) was considered for measurement and correction, so that only the first 3 joints have to be considered. To simulate the real behavior of the manipulator, we have introduced backlash, gear runout and joint flexibility.

For the training stage of the neural network, 80 measurement lines, all parallel to the y_0-axis, were selected. These lines are inside the largest cube contained in the spherical workspace of this robot. Each line has a length of 0.4 m with 9 different measurement points located on it. During the training, the end effector is placed on each of the 720 points and the difference between the real position and the nominal one is measured. For each point, two different payloads are considered (M=0kg and M=1kg). The total number of measures for the training set is therefore 1440. Each element of the training set consists of an input pattern specifying the angular configuration and payload of the robot and of an output pattern giving the positioning error. Once the training stage is finished, a validation phase is performed in order to test and to quantify the performances of the neural network. For this, we have introduced several validation lines parallel to the training lines and inside the reference cube, and other lines with payloads different from the training ones to assess the interpolation and generalization capabilities of the network. The number of elements used for the interpolation and generalization test are 64 and 36, respectively.

Figure 6 (dashed line) gives a typical positioning error profile along a "demonstration" line, perpendicular to the training lines. Effects of runout (oscillations) and backlash (discontinuity) are clearly observed.

Before any correction, the mean positioning error (r.m.s.) measured over the training set is about 1.0mm. Note that the mean value of the error (r.m.s.) is defined by :

$$\bar{e} = \sqrt{\frac{1}{N} \sum_{m=1}^{N} \sum_{i=1}^{3} e_{m,i}^2}$$

where N is the number of measurement points and $e_{m,i}$ is the ith component of the positioning error at the mth point.

The procedure described in section III was applied for developing a good neural network architecture. The resulting structure was composed of two hidden layers with respectively 16 and 4 units. After learning, the mean estimation error (difference between the real positioning error and the output of the network) was of the order of 0.2mm, both on the training lines and on the validation lines, as illustrated by Table 2 (second column). For these cases, the network shows effectively good interpolation and generalization capabilities. Figure 6 (continous line) gives the estimated positioning error along the "demonstration" line; it can be observed that the network has considered runout effects as noise and has smoothed over the real profile. The introduction of auxiliary inputs (such as $\cos(n\theta_i)$ and $\sin(n\theta_i)$, where n is the reduction ratio of the gears) is likely to help the effective learning of this kind of effect.

Table 2 - Mean Errors for Training Set and Test Set

	Positioning Error Without Correction	Estimation Error	Positioning Error After Correction
Training Set (1440 points)	1.10 mm	0.22 mm	0.24 mm
Test Set, Generalization (36 points)	1.08 mm	0.22 mm	0.24 mm
Test Set, Interpolation (64 points)	1.07 mm	0.23 mm	0.25 mm

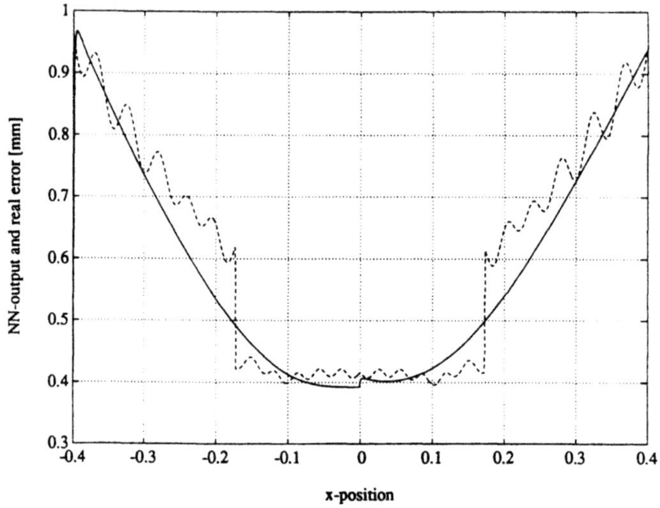

Fig.6. Comparison of NN-predicted error and real position error

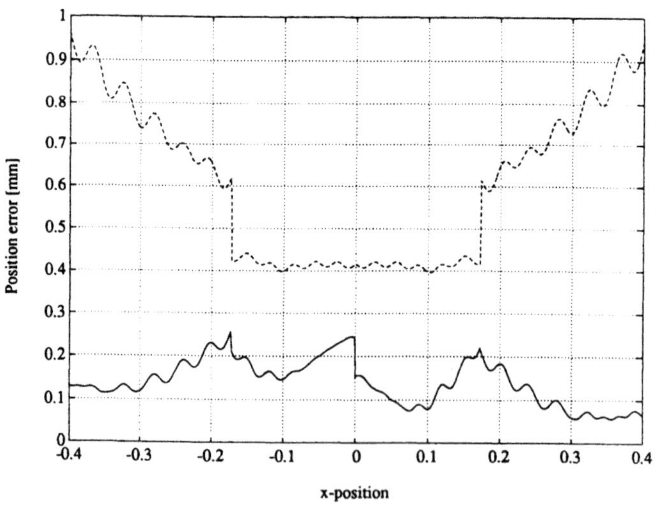

Fig.7. Positioning error before and after calibration

Finally, we have implemented the two compensation techniques presented in section 4. Both have been tested on the training lines and the validation lines, and give very similar results, as expected from eq. 6 and eq. 9. Figure 7 presents the positioning error before and after correction along the demonstration line. The third column of table 2 gives the r.m.s. positioning error after compensation following the false target technique. Comparison between column 1 and 3 shows that an improvement of a factor 5 in the accuracy can be achieved by the present technique. On the other hand, comparison between column 2 and 3 shows that the remaining positioning error after correction is mainly due to the estimation error of the neural network.

6. Conclusions of Part II

A new technique for the calibration of robot based on a neural network approach for the identification of non-geometrical errors was presented. After learning of the network and correction by a false target technique, the positioning error can be reduced up to 20% of the initial error. Although the neural network seems to have sufficiently good generalization and interpolation capabilities, it fails to discover and to learn high-frequency effects such as runout. Further work must focus on this problem (for example, by considering the introduction of auxiliary inputs to the network) and address real experiments on a physical robot.

This technique, in combination with the first one presented in part I, should provide an efficient mean to reduce the positioning error by a factor of more than 50 (2% of the error before any calibration).The fact that a lower bound exists on the calibration error, dictated by robot repeatability (quasi-stochastic effects) and accuracy of the measurement system, should be kept in mind. Nevertheless, with this level of performance, off-line programming of robotics systems can become an efficient, safe and reliable tool in the field of remote handling and teleoperation.

ACKNOWLEDGMENT

The authors would like to thank Dr. M. Becquet, J. del R. Millan, N. Mitchison from the Joint Research Center of Ispra (Italy), Pr. R. Hanus, Dr. M. Kinnaert and E. Rossignol from the Free University of Brussels for their helpful comments and suggestions.

REFERENCES

[1] W.K. Veitschegger and C.H. Wu (1987). A Method for Calibrating and Compensating Robot Kinematic Errors, *Proc. IEEE Int. Conf. Robotics and Automation 1987*, 39-44.

[2] M.-S. Cai and A. Rovetta (1988). Position and Orientation Error Analysis and a Contribution to Newton's Method for Solving the Inverse Kinematic for Robot Manipulator, *Proc. International Conference on Computational Intelligence 1988*, 211-220.

[3] J.M. Hollerbach and Bennett (1988). Automatic Kinematic Calibration using a Motion Tracking System, *Robotics Research: the Fourth International Symposium*, MIT Press, 191-198.

[4] C.H.An, C.G. Atkeson and J.M. Hollerbach (1988). *Model-based Control of a Robot Manipulator*, MIT Press, Cambridge, 49-64.

[5] A. Preumont, E. Ponslet and D. Lefeber (1989). Improvement of the geometrical model of robot manipulators, *Proc. of the Int. Symposium on ACD & D 1989.*

[6] B.C. Jiang, J.T. Black and J.N. Hool (1989). Taguchi Methods-based Robot Metrology, *Proc. Robots 13 Conference*, 145-159.

[7] M. Becquet, J.M. Renders and E. Ponslet (1988). Kinematic model sensitivity to geometrical parameters errors and its impact on robotic precision, Proc. *Third International Conference on CAD/CAM, Robotics and Factories of the future*, 160-165.

[8] M. Becquet and D'Haemers (1988). Static and Dynamic Performance Measurements of Industrial Robots, *Revue M*, **32**.

[9] D. Payannet, M.J. Aldon and A. Liégeois (1985). Identification and Compensation of Mechanical Errors for Industrial Robots, *Proc. 15th ISIR*, 857-864.

[10] M. Ishii, S. Sakane, M. Kakikura and Y. Mikami (1986). A new Calibration System for Improving Absolute Accuracy of Robot Manipulators, *Proc. 16th ISIR*, 1017-1025.

[11] J. Chen and L.M. Chao (1986). Positioning Error Analysis for Robot Manipulators with all Rotary Joints, *Proc. IEEE Int. Conf. Robotics and Automation*, 1011-1016.

[12] M. Ishii, S. Sakane and Y. Mikami (1987). A 3-D sensor system for Teaching Robot Paths and Environments, *The International Journal of Robotics Research*, **6**, 45-56

[13] D.E. Whitney, C.A. Lozinski and J.M. Rourke (1986). Industrial Robot Forward calibration Method and Results, *ASME Journal of Dynamic System, Measurements and Control*, **108**, 1-8

[14] R.P. Judd and A.B. Knasinski (1987). A Technique to calibrate Industrial Robots with Experimental Verification, *Proc. IEEE Int. Conf. Robotics and Automation, 1987*, 351-357.

[15] H.W. Stone and A.S. Sanderson (1987). A Prototype Arm signature Identification system, *Proc. IEEE, Int. Conf. Robotics and Automation, 1987*, 175-182.

[16] J.H. Gilby and G.A. Parker (1982), Laser Tracking system to Measure Robot Arm Performance, *Sensor review* **2**(4), 180-184.

[17] M. Becquet (1987). Analysis of Flexibility sources in Robot structures, *Proc. IMACS/IFAC Int. Symp. on Modeling and Simulation of Distributed Parameter Systems 1987*, 419-424.

[18] W.K. Veitschegger and C.H. Wu (1986). Robot Accuracy Analysis Based on Kinematics, *IEEE Journal of Robotics and Automation*, **RA-2**, 171-180.

[19] L. Ljung (1987). *System Identification - Theory for the user*, Prentice-Hall (New-Jersey), 1987

[20] C.H. Menq and J.H. Borm (1988). Statistical measure and characterization of robot errors, *Proc. IEEE Int. Conf. Robotics and Automation, 1988*, 926-930

[21] Z.S. Roth, B.W. Mooring and B. Ravani (1987). An Overview of Robot Calibration, *IEEE Journal of Robotics and Automation*, **RA-3**, 377-384

[22] D.E. Rumelhart, G.E. Hinton and R.J. Williams (1986). Learning representations by back-propagating errors, *Nature*, **323**, 533-536.

[23] R. Hecht-Nielsen (1990). Theory of the backpropagation neural network, *Proc. Int. Joint Conf. on Neural Networks, Vol. I*, 593-611.

[24] D.C. Plaut, S.J. Nowlan and G.E. Hinton (1986). Experiments on learning by back propagation, Technical Report CMU-CS-86-126, Carnegie-Mellon University.

[25] J.M. Renders, J. del R. Millan, M. Becquet (1991). Non-geometrical Parameters Identification For Robot Kinematic Calibration by Use of Neural Network Techniques, *Proceedings of the European Robotics and Intelligent Systems Conference*, Kluwer Acadamedic Editors (to appear).

TRANSPORTERS FOR TELEOPERATIONS IN JET

T. RAIMONDI
JET Joint Undertaking
Abingdon
Oxfordshire
OX14 3EA

ABSTRACT In order to carry out the remote operations that are envisaged, it must be possible to position a dexterous manipulator in the working area inside and outside the vacuum vessel. Owing to the size of the machine the transporters designed to do this have to have a very large reach. At the same time, extremely precise positioning is required as access into the vessel is by way of ports of small dimensions, and space around is cluttered with components and services. So the transporters themselves assume the features of a very large robotic device. The control system, microprocessor based, includes teach-and-repeat and resolved motion facilities.

1. Introduction

In order to carry out the remote operations that are envisaged, it must be possible to position a dexterous manipulator in the working area inside or outside the vacuum vessel. Owing to the size of the machine the transporters designed to do this have to have a very large reach. At the same time, extremely precise positioning is required as access into the vessel is by way of ports of small dimensions, and the vast array of peripherals surrounding it constitute obstacles that must be avoided. So the transporters themselves assume the features of very large robotic devices.

2. Articulated Boom For In-Vessel Operations

As there are some heavy components that must be replaced periodically, the in-vessel transporter also serves to install and dismount these and to convey them in and out of the vessel. These operations require high

M. C. Becquet (ed.), Teleoperation: Numerical Simulation and Experimental Validation, 87–107.

precision and repeatability and also compliance for the final homing of the component onto its fixtures. For this purpose the manipulator can be removed and replaced by a series of ad hoc end effectors to handle these components. A single action quick coupling has been designed to facilitate the changeover. As the mechanical connection is made the electric, water and gas services are connected automatically.

2.1 LAYOUT

In a tokamak like JET the only convenient access to the torus is through ports situated on the equatorial plane between the toroidal coils. The articulated boom enters the vessel through either of two opposite horizontal ports. It can reach round half the circumference with a payload of one tonne. With the addition of another segment the reach is extended right round to the opposite port, 14 metres from the support trolley, with a payload of 450kg (Fig 1). The alternative of a transporter running on tracks was discarded because of the difficulty of getting it through the port and because any fixtures inside the vessel would be susceptible to plasma damage. Umbilical cables would also pose an extremely difficult problem. The geometry is unsuitable for a telescopic boom.

The boom consists of five segments cantilevered off a trolley suspended from a beam which spans the gap between the wall of the torus hall and the machine. To compensate for misalignments and deflections due to the load (max 70mm at full extension), the trolley allows for limited slew and tilt. The end segment, known as the boom extension, can be tilted to position the end effector close to the top or bottom of the vessel. It has pan/tilt/roll motions and terminates with the quick connector. One section of the boom extension is an open structure providing a receptacle for tools to be used by the manipulator (Figs 2, 3).

2.2 STRUCTURE

Care was taken in the choice of materials to optimize strength vs weight. The first two segments are in stainless steel and the others in aluminium alloy. These are all closed box structures to resists torsion. The boom extension is an aluminium/magnesium casting, strengthened by hot isostatic pressing. It had to be an open structure to accommodate the tool box and was optimized with FE calculations. [1]

2.3 ACTUATORS

Hydraulic fluids were excluded by the high vacuum requirements so all the motors are electric. The actuators are on the joints (Fig 4) and there is space above them for the cabling to be routed between the top and bottom hinges.

Fig. 1 Articulated Boom in Mock-up Area.

Fig. 2 Boom Extension with T.V. Arms and an Antenna Grabber.

Fig. 3 Boom Positioning a Limiter Segment in the Vessel.

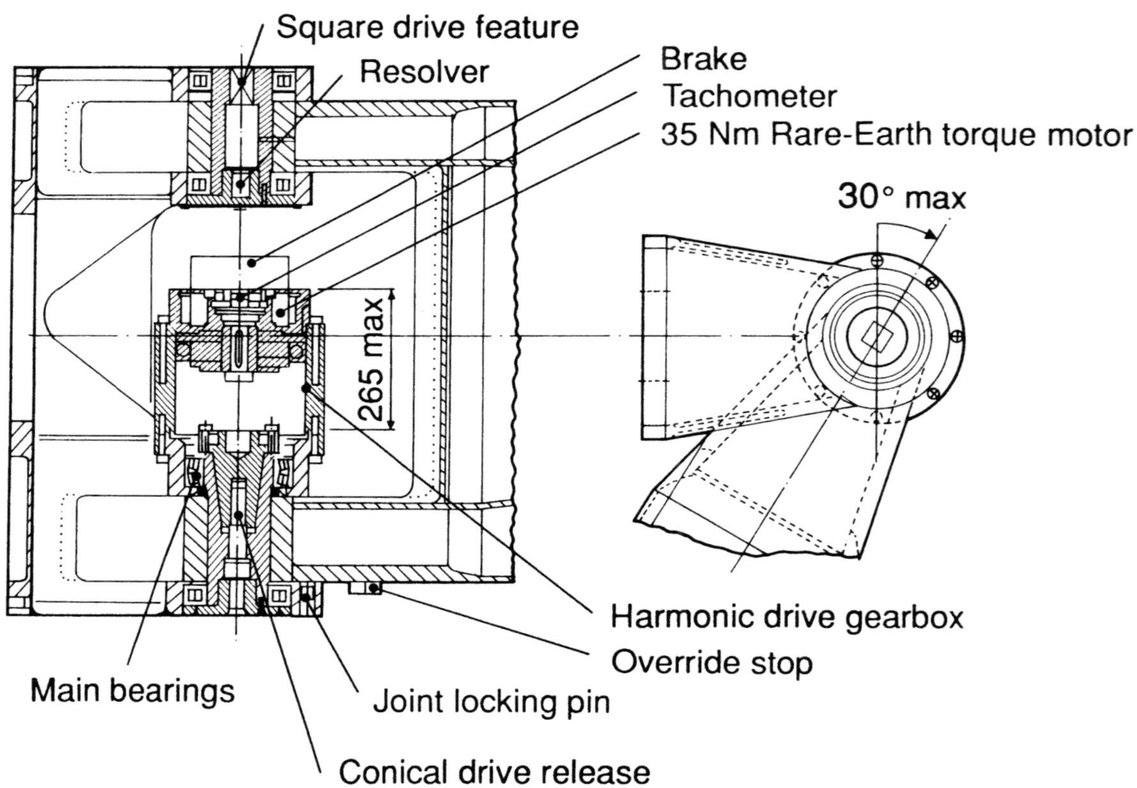

Fig. 4 Actuator.

The actuators consist of Harmonic Drive gearboxes, directly coupled to Inland DC torque motors, incorporating a DC tachogenerator and a brake. This solution offers the desired compactness for the high torque required (5000Nm), fast response and negligible backlash. This in turn permits installation directly on the joint of a resolver, coupled to a 16 bit R/D converter, which gives the required repeatability of 1mm.

The main shortcoming of the Harmonic Drive gear is a certain degree of elasticity. On the other hand some passive compliance can be beneficial as the boom comes into collision with the wall.

The torque is transmitted from the Harmonic Drive to the segment by means of a torque arm which engages the front plate of the segment in a precision machined slot. This allows for certain degree of freedom, thus decoupling the axis of the Harmonic Drive from the axis of rotation of the segment. Spherical roller bearings are used for the hinges to avoid misalignments. In the event of a major failure it could be necessary to retrieve the boom from the vessel using a manipulator mounted on the TARM. The torque arm can be disengaged from its slot and the segment is free to rotate.

2.4 CONTROL

The rotation of each of the main joints and the translation of the support trolley are controlled with a servo loop configuration as in Fig 5. The position error, calculated via software, is applied as an input signal to hardwired velocity servos controlling the motors with PWM (pulse width modulation) amplifiers. The torque T developed by each gearbox depends on the angular difference between its input and output shafts and is a function of its elasticity, backlash and efficiency in both direct and backdriven modes. The single joint servo loops are reciprocally coupled through the system dynamics which virtually corresponds to the inertia matrix of the five vertical joints.

The architecture of the control system is shown in Fig 6. The following functions are provided: [2]

- Closed loop control of all joints and open loop control of adjustment jacks and various end effector motors.

- Control of the boom from either the Remote Handling Workstation or a local handbox.

- Safety interlocks to prevent joint runaway or servo errors.

There are three modes of operation:

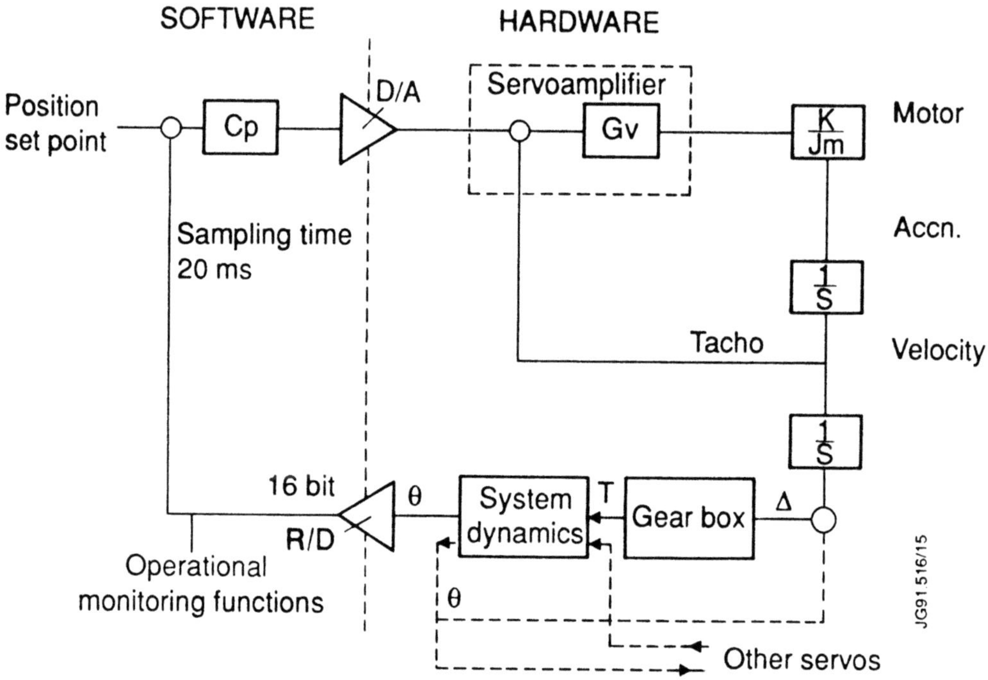

Fig. 5 Servoloop of One Joint.

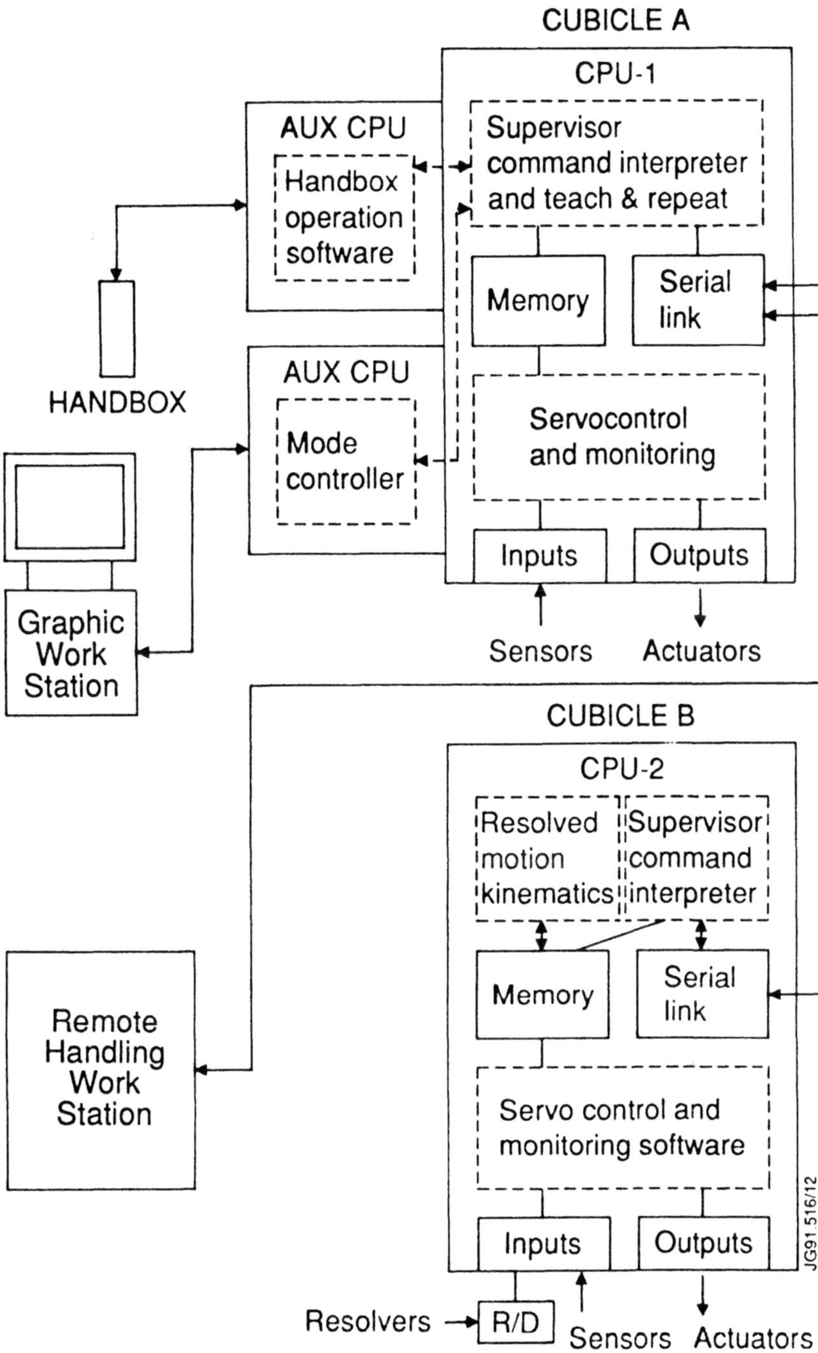

Fig. 6 Architecture of Control System.

a) Direct mode - one or two joints can be selected and moved in rate control from a joystick or position controlled from a keyboard.

b) Resolved motion - either in "tool" co-ordinate frame located on a preselected segment or "world" co-ordinate frame in fixed position.

c) Teach-and-repeat, which allows previously taught sequences to be automatically repeated.

The operator interacts with the systems via:

- A Remote Handling Workstation, provided with touch screen to select allocation of joints to joystick, rocker switches allocated to various functions, proportional 3 d.o.f joystick and 2 colour monitors.

- A small portable handbox, mainly to be used for commissioning, with 3 d.o.f joystick and keyboard.

- A Graphics Workstation which offers a powerful visual aid, showing in real time the position of the boom relative to the environment. It is a substitute for the camera when the latter cannot provide a good view, and also offers a facility for off-line teaching.

2.5 DYNAMICS

Due to the exceptional dimensions, the inertias of the boom and the elasticity of the joints, dynamic cross-coupling between the joints has to be taken into account. A dynamic model has been set up and validated.[3] The actuator was modelled as in Fig 7. The elasticity of the actuator was concentrated at the output shaft (stiffness Hi).

Fig 8 shows the block diagram of the servosystem in its present configuration. The input signal θiR is the demand of position of the segment i, relative to the segment i-1. The position error, amplified by an adjustable parameter GPi, is the demand of relative velocity. The velocity error is transformed through the current amplifier K into electromagnetic torque applied to the motor. This is summed to the torque feed-back applied by the fast shaft of the gearbox on the motor, which depends on the load. It is affected by the efficiency of the gearbox which in turn depends on the flow of the mechanical power. In the direct driven mode, when the torque applied to the load has the same sign as the velocity (power delivered by the load), the torque Ti/ri is supplied through the block defined by $1/\eta_d$ where η_d is the efficiency of the gearbox in the direct mode, assumed as 0.8. In the back-driven mode, when torque to the load and velocity have opposite sign (power delivered to the gearbox by the load) the

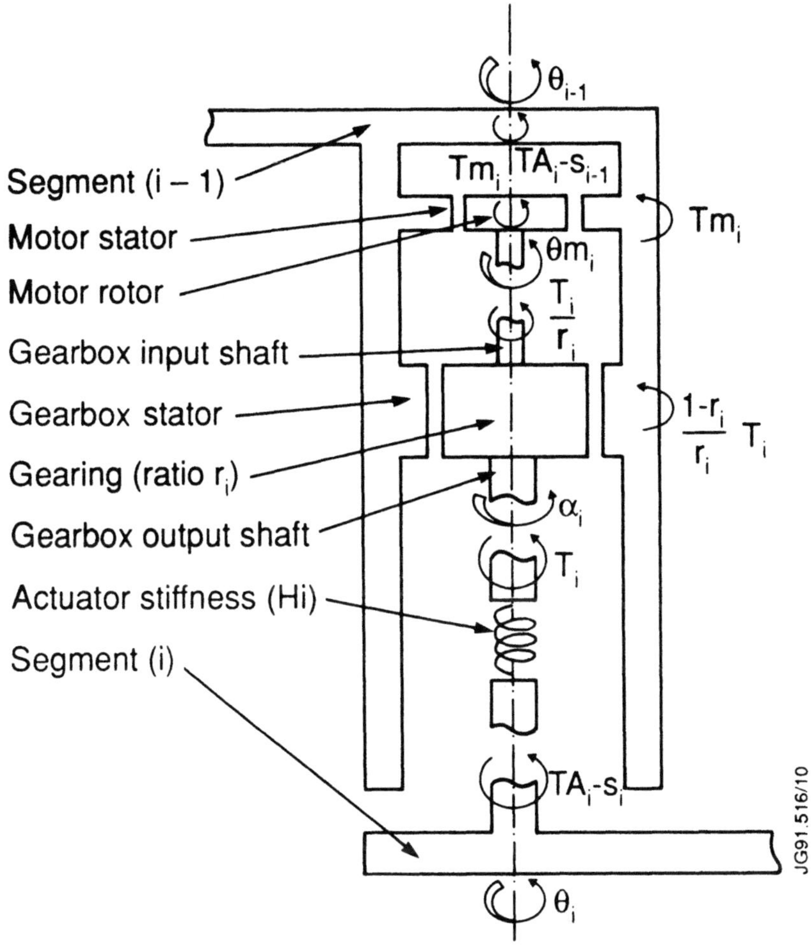

Fig. 7 Model of Actuator.

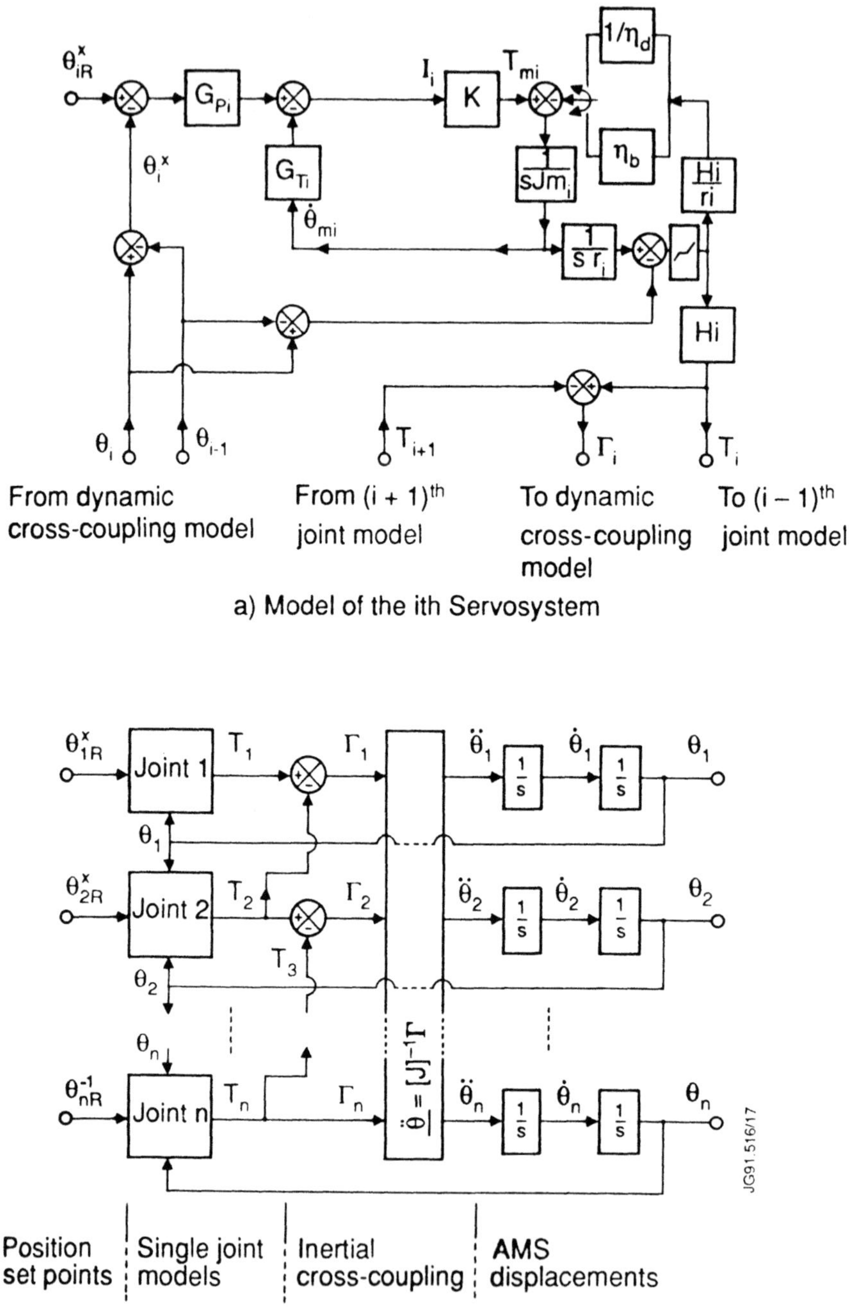

a) Model of the ith Servosystem

b) Model of the Cross-coupling

Fig. 8 Model of Servosystem.

torque Ti/ri is supplied through the block defined by η_b, where η_b is the efficiency in the backdriven mode assumed as 0.6. The resultant torque to the motor is integrated to give the absolute velocity and position of the motor. The velocity signal is used as an analogue feed-back, neglecting the small effect of $\dot{\theta}_{i-1}$. The position is used to evaluate the torsional angle between input and output of the gearbox, which after a deadband simulating the backlash, assumed at 3' of an arc, is multiplied by the stiffness Hi algebraically summed to the torque Ti + 1 at the other end of the segment to give the overall torque Gi. The vector $|\Gamma|$ is then multiplied by the inverse of the inertia matrix to give the vector $|\ddot{\theta}|$ which, integrated twice, gives the vector position $|\theta|$ used as a feed-back. Since the boom moves very slowly Coriolis and centripetal effects have been ignored. A comparison of step responses between the real boom and the model is shown in Fig 9.

With the simple position servosystem described, the boom tends to have an oscillatory response. The lowest frequency of the oscillation modes is in the order of 0.3Hz, which is too close to the frequency cut-off of approximately 0.2Hz. To avoid overshoots and to have good trajectory control, the position demand signal must vary very gradually. The operator using a joystick does not have a rapid enough perception of the response of the boom and tends to overreact. In teach-and-repeat the input signals can be very smooth at very low speeds.

2.6 OPERATION OF TRANSPORTERS

2.6.1 *Prepositioning Phase.* In JET navigating the articulated boom to position end effectors, manipulator and cameras within the vessel is an arduous task for the operator using joystick and TV because of the labyrinth of services around the machine and the narrow entrance port. The problem is resolved using teach-and-repeat files worked out on the graphics model and verified and adjusted in the real environment. For a future reactor, in spite of the larger dimensions, similar difficulties are likely to arise. The operation of teaching the transporter movements on the graphics model should be performed during the design phase as part of the configuration control. This activity will ensure the access space vital for maintenance and also be useful for the first installation.

2.6.2 *Final Positioning of End Effectors.* Homing of end effectors on the machine requires final adjustment with visual feed-back owing to movements of the vessel, alterations to the vessel surface during operation or repeatibility errors of the transporters. Although homing can, in principle, be done with joystick and TV viewing, in practice it is made difficult by the inertia of the load and imperfect viewing. A better method could be to prepare to teach-and-repeat file right to the end position and

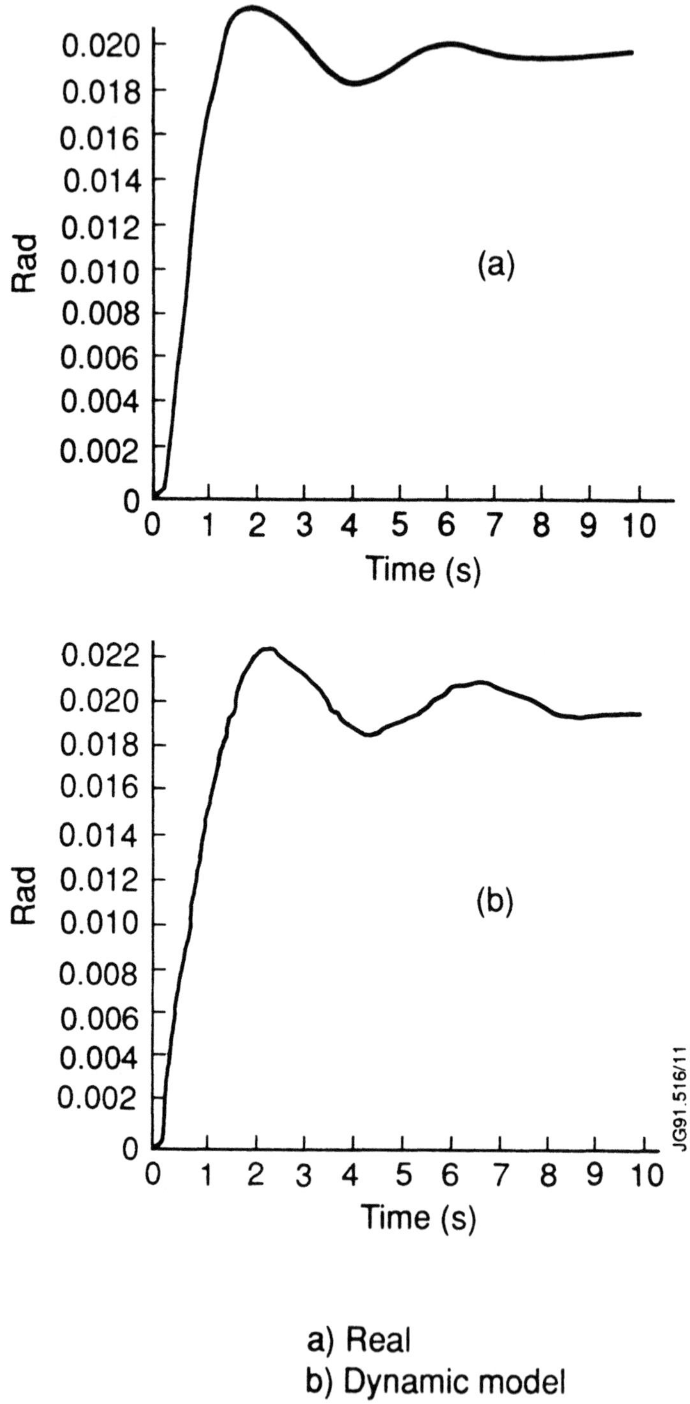

Fig. 9 Step Response.

apply corrective cartesian offsets at suitable visual checkpoints. From experience with the JET boom it appears that homing accuracy of about 2mm is achievable at low speed. The passive compliance of the system compensates this final error. At present a teach-and-repeat file for positioning an antenna from pick-up point outside the vessel to final position takes about 7 mins. With the backlash-free actuators, the stick-slip effect is not a problem and dynamic tracking accuracy in the region of 5mm has been achieved by imposing gradual accelerations. Torque loops could in principle be added to the boom actuators in order to achieve active compliance and improve the bandwidth of the position servos. Modelling of this system at JET shows that the overall speed could be improved by a factor of 3. Taking account of the safety factor, it is doubtful whether it is a worthwhile gain.

2.6.3 *Safety Studies.* Accurate fault analysis of remote handling devices, retrieval strategies in-emergency and obstacle avoidance methods are an essential programme. A collaboration has been established between JET and CCR Ispra for safety studies of the JET equipment. These should be of interest and value for the NET and ITER designers. The principle of an obstacle avoidance system using TV cameras and image processing has been tested in JET. An attractive alternative would be a system of collision warning built into a real time simulator.

3. Ex-Vessel Transporters

The TARM (Telescopic Articulated Remote Mast) is the main transporter for access to areas on the outside of the vacuum vessel. It will be suspended from the crab of the main 150 tonne crane, which was used for the installation of the machine. Auxiliary transporters are the low-level transporter and the turret truck.

3.1 MAIN 150 TONNE CRANE

The crane was specified with the fine controls that would be required during remote operations. This turned out to be a bonus during the assembly phase. Positioning the 130 tonne octants with tolerances of about 1mm was no problem. A trial of inserting one octant with TV viewing was successful. Minimum incremental displacements of the load were of the order of 0.2mm in the vertical direction and 1mm horizontally, with negligible swinging effects. All the crane motors are thyristor controlled. The load is continuously monitored by means of load cells. Given the elasticity of the ropes and the low controllable speeds, vertical contact loading can be kept below 300kg. The rotation of the 150 tonne ramshorn hook is motorized. This hook will be replaced by a shackle for remote engagement to lifting eyes. The mechanical repeatability of the position of the hook in any co-ordinate is 20mm. To lift large, delicate structures such as the outer poloidal field coils, a four-rope system with hydraulic load equalizers is used. The system will also be used to raise to its location under the crab the structure which will support the telescopic arm (TARM).

3.2 CRANE-MOUNTED TELESCOPIC ARM (TARM)

The TARM will be attached to the crane via the 4-rope system originally specified for lifting the large poloidal coils. Fig 10 shows a schematic of the TARM motions. All major axes are designed with zero backlash drive units. Fig 11 shows the operating configurations required to enable the Mascot to access positions not only above, below, and down the central core of the torus, but also halfway across inside the vacuum vessel in order to work in conjunction with the articulated boom during installation of heavy components.

A 250kg jib crane (crane K, Fig 10) is positioned on the moving vertical telescope so as to be able to lift very long diagnostic equipment out of the top of the machine. A 6 tonne jib crane with motorized hook will carry out most lifting operations in conjunction with the Mascot.

The vertical mast can be coupled directly to the Mascot for access down the middle of the machine or the horizontal telescope may be attached with a purpose designed remote attachment system. A limited (+10, -15) tilt action (A1) allows for deflection compensation and also allows the Mascot to reach 0.9m below the floor level for basement maintenance.

In order to minimize cable requirements and for ease of commissioning, the service modules are all attached to the upper 'fixed' part of the TARM. These include power supplies and local controllers for the drive units, signal conditioning for sensors and TV cameras, Mascot slave controller, water supplies for tools, air supply for tools and air make-up for gaiters.

The microprocessor based control system will have similar features to that of the boom, with single joint closed loop servo control, teach-and-repeat and resolved motion. Here also, control will be from the Remote Handling Workstation, from a portable handbox, or from the Graphics Workstation.

3.3 AUXILIARY TRANSPORTERS

Some components on the lower half of the vacuum vessel which will require handling are not accessible to the overhead crane. For this we have the low-level transporter, which is a remotely controlled modified electric fork-lift truck equipped with an attachment and end effector capable of precisely controlled movements. It will be used in conjunction with a

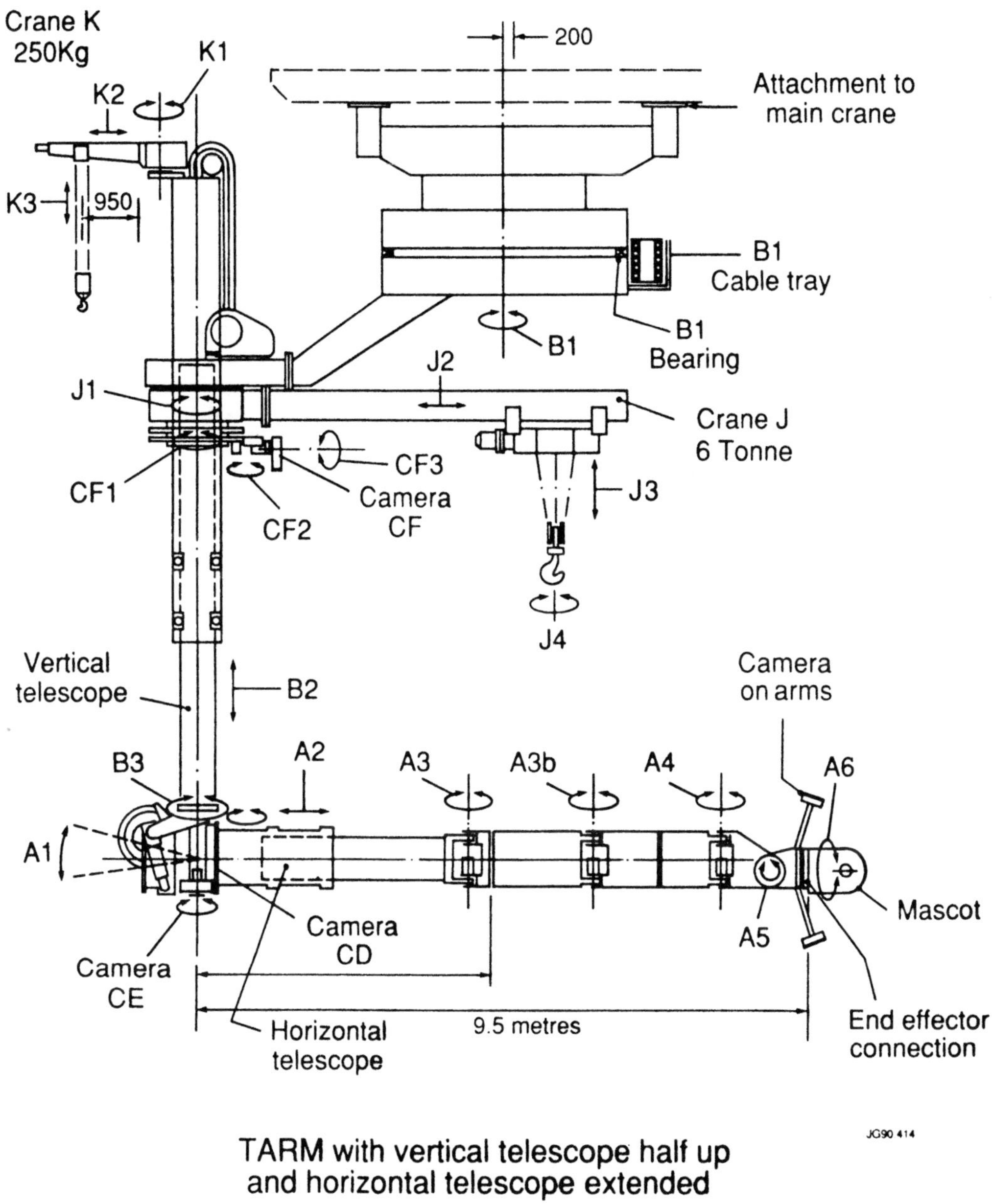

Fig. 10 TARM Movements.

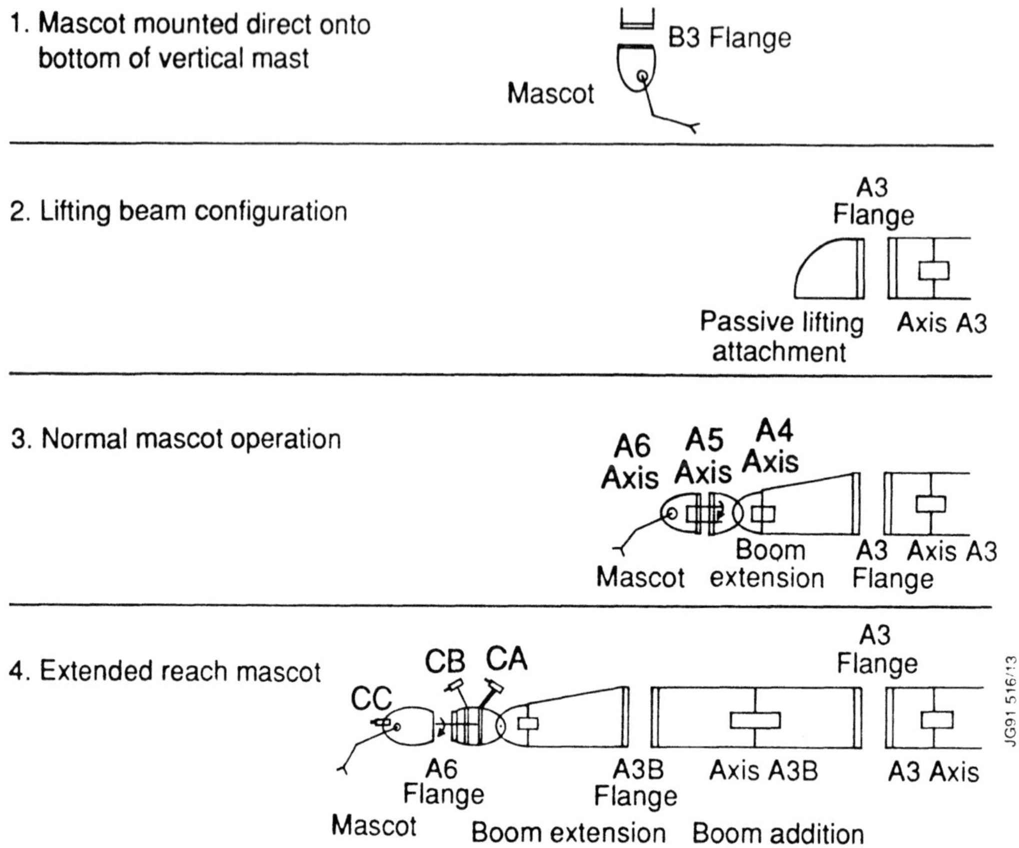

Fig. 11 TARM Configurations.

servomanipulator supported by the TARM. It was used to install several TM pumps and VAT valves (Fig 12).

Acknowledgments

Thanks are extended to all the members of the Remote Handling groups, whose work forms the basis of this paper

Fig. 12 TARM installed on stillage in JET, with Mascot approaching a turbopump on a low level transporter for Remote Handling trials. Both TARM and boom can reach the inside of the spare octant for in-vessel tests.

References

[1] L.P.D.F. Jones, D. Maisonnier, T. Raimondi, Design and Operation of the JET Articulated Boom, Proc. 11th S.O.F.E., Austin 1985.

[2] L. Galbiati, T. Raimondi, Control and Operation of JET Articulated Boom, Proc. 12th S.O.F.E., Monterey 1987.

[3] D. Maisonnier, Modelling and Control of a Large Robot for Teleoperation System, Proc. of 1987 I.E.E.E.

[4] L.P.D.F. Jones, L. Galbiati, M. Gredel, W. Neddermeiyer, The Design and Construction of the TARM - a Crane Mounted Remotely Controlled Transporter for JET.

NUCLEAR TELEOPERATION
PARTICULAR CHALLENGES IN DECOMMISSIONING APPLICATIONS

Marc Decréton
CEN/SCK Nuclear Research Centre
Boeretang, 200
B-2400 Mol
Belgium

ABSTRACT. Decommissioning activities sets particular requirements to teleoperated equipments. Advanced features are needed to cope with highly unstructured and unpredictable environments. On the other hand, high reliability specifications must be reached, taking into account the severe conditions on radiation and contamination levels. A survey is presented of the typical tasks to be expected, as well as the particular requirements and challenges put on advanced teleoperated equipments to be used in decommissioning applications.

1. Introduction

Many of the usual operation and maintenance tasks in reactors and fuel processing facilities have been designed for manned entry. Operating experience has shown however that more and more interventions, especially for repair, decontamination and decommissioning work, present a high level of radiation and contamination risk. This has become even more accute as current practice is now to keep exposure "as low as reasonably achievable" (ALARA). In analysing the radiation protection and trying to optimise it, remote operation appears naturally as a key factor. However, traditional telemanipulators have such limitations, in reach, in payload, in dexterity, that many complex tasks cannot be envisaged. Advanced teleoperation, incorporating a high level of both autonomy and intelligence, as well as mobility and broader reach capabilities, are necessary. But in order to get really a positive radioprotection balance, these equipment must offer high reliability, and some kind of self or remote rescuing possibilities, in case of failure. Extensive man entry needed to maintain robotised equipment would jeopardise any potential benefit of such techniques.

M. C. Becquet (ed.), Teleoperation: Numerical Simulation and Experimental Validation, 109–138.

Another rationale for more automation in the nuclear sector, is linked to the more traditional arguments valid in any industrial process. Reduction in personnel costs, lower down times, more reproducibility, reduction in human errors alias, are typical parameters. Such arguments are mainly pertinent to manufacturing tasks as they mainly appear in fuel processing facilities, but can be applied to many other domains of the nuclear industry, including decommissioning applications. The main difference, and constraint, here is that most systems have not been built to take into account any remote operation, and that the requirements put on equipment, w.r.t. flexibility and autonomy, are critical.

These general considerations are also valid for the particular case of decommissioning. However particular characteristics must be considered here:

- Decommissioning covers a wide range of applications, ranging from tasks similar to industrial manufacturing, to specific demanding situations related to dismantling environments.
- Any application is unique. Machines are not built for long term fabrication processes or multiple maintenance cycles.
- Environmental conditions are severe: high radiation and contamination risk, poor geometrical definition of the environment, underwater work in some cases, etc.

In considering advance teleoperation for decommissioning applications, one should keep in mind some basic engineeering principles:

- The solutions must be kept **simple**, but the simplicity must be considered globally. The use of simple long tools instead of more versatile manipulators in complex tasks requires for instance many more operators, and transfers therefore the complexity to the training and management aspects.
- Systems must be **flexible** enough to cope with unpredictable situations. Dismantling is basically a process generating uncertainties, partially non-deterministic in its evolution.
- **Generic** solutions are more effective than many particular systems, when training, maintenance and adaptation to unforeseen situations are considered.
- **Cost** must be considered as a key decision parameter, but on a global scale, i.e. taking into account not only equipment cost, but also personnel dose management, training, down times, etc.

This paper will present first a survey of typical nuclear applications of intelligent teleoperation, as well as of the different classes of machines used or in development. It will then detail more specifically the different tasks to be expected in any decommissioning activity. Finally the main issues and challenges of these applications w.r.t. present technology, are presented.

2. Nuclear applications of intelligent teleoperation

Intelligent teleoperated systems are beginning to be used and will experience a certain important growth in virtually all major sectors of the nuclear industry (1). In particular, the following areas can be cited.

2.1 Reactor operation and inspection

This area has seen already some important applications of intelligent teleoperation; however, mainly on dedicated machines, built to perform a specific task, as for instance the refueling operation or the steam generator inspection (2), but featuring more and more advanced automated sequences, improving the safety, the human interventions and the reactor down-time (3). Other systems are in the design and test phase. They are mainly concerned with weld and leak inspection (4), decontamination and clean-up, radiation/-contamination and other key parameters monitoring, waste and fuel handling, general remote viewing systems (5)(6)(7)(86)(88). Specific attention is put on transportation mobile platforms both suitable for general access, and for specific positions such as for weld or leak control (8)(9)(10)(11)(92).

2.2 Reactor maintenance

Reactor maintenance operation require particular operations, such as welding, mechanical fastening, cutting tasks, mechanical and electrical replacement of components (heaters, filters, detectors, valves, drives, etc.), pump maintenance work, pipe cutting, welding and cleaning, fuel and waste manipulations (12)(87). Some existing systems are in operation, as for instance some steam generator maintenance machines (2)(17)(18), pipe crawlers (13)(23)(93), etc. However, general purpose teleoperated systems are still under development, as it is the case for inspection (14)(15)(16)(20).

2.3 Fuel fabrication and reprocessing

Utilisation of telemanipulators for fuel operations in cells (fabrication, reprocessing, examination) are well established. The introduction of specific automated features is presently more and more introduced in new designed facilities (19). These types of applications have the advantage to be quite similar to many industrial fabrication processes, and can make profit of the rapid developments occurring in the domain of computer aided manufacturing (21). The nuclear implementation requires naturally specific characteristics such as radiation and contamination tolerance, remote access and usually less structured work environment (22). Many applications are already.operational in many fuel handling plants, for fuel material handling and managing, pin loading, decontamination processes, chemistry analysis, furnace maintenance, etc (24).

2.4 Waste handling

Waste handling is usually one of the most appropriate operations to be automated using current industrial practice, as the tasks are very similar to the ones appearing in many industrial factories (25). Typical examples are waste encapsulation and storage, filter replacement, automated laundry, handling of waste containers and monitoring, waste inventory control and storage area surveillance, samples analysis, etc. Some concrete applications have found their way, based sometimes on existing industrial robots (26)(27).

2.5 Dismantling

Decommissioning of nuclear facilities involve a broad range of activities requiring the ability to dismantle major equipments in the presence of radiation field and contamination. In some instance, tasks have to be performed under water. Typical applications can be seen in the following domains.

- Survey: Components to be removed must be surveyed to assess their degree of contamination and activity, as well as determining the levels of radioactivity and contamination in the surrounding potential work area.
- Decontamination: components or work areas must be decontaminated prior to dismantling work.
- Dismantling: This work involves the removal of components from the facility, as well as in-situ (or in special work-cells) cutting oversize components in smaller pieces. Particular attention must be put here on contamination spread avoidance and secondary waste collection.
- Waste packaging: processed radioactive and contaminated items (resulting from removal and cutting operation, from secondary waste collecting devices, or any tools that has been used for the given tasks) must be transported to a dedicated packaging area, processed in containers, and stored.

As this paper will focus in more details on the tasks and requirements related to decommissioning work, these aspects will be presented in more details in the next chapters.

2.6 Post-accidental intervention

Experience in using teleoperated equipment following serious accidents is fortunately limited. However, their need has been acutely felt, both at TMI and Chernobyl. Limited assessment of specific machines was achieved, not always with satisfactory results. Due to the nature itself of partially unpredictable situations, it is here difficult to list the different tasks to be expected, and requirements on versatility and flexibility are the most important (30)(32).

The first type of tasks is related to problem assessment. The scale of the accident must be determined, remotely, and eventually under arduous conditions, w.r.t. access capabilities and radiation exposure. Data must be obtained on radiological release, condition of main equipments, characteristics of the dispersed materials. Machines to operate in such conditions should not only be able to tolerate the harsh environment and perform the necessary measurements, but would also be capable to free access routes in some extend (28).
In a second phase, sometimes called stabilisation phase, remote equipments are needed to avoid any worsening of the situation. Leakages should be stopped by manipulating valves, preparing dams, bunds or any needed coverage (29)(31).
The third phase is related to the rehabilitation of the site, and is in some aspect similar to the decommissioning problematics, but with particular difficulties. As normal dismantling starts from a known and stable structure, post-accidental environment is highly unpredictable, unstable and unstructured. On the other hand, large scale contamination can be present, and clean-up operation is a critical part of the work (36) (ejected debris collection, removal of contaminated soil or masonry, pumping of radioactive liquids, etc). Post-accidental situations are clearly the most demanding, as detailed requirements cannot be formulated in advance, work in completely unstructured environment is needed, and eventually very high radiation doses and contamination.

3. Types of machines used in nuclear teleoperation

From the beginning of nuclear industry, remote operation has been a critical issue for handling highly radioactive or contaminated material. One can divide the different approaches in the following categories: dedicated machines, custom-made for a given specific task, master/slave manipulators, mainly used for cell work, advanced telemanipulators where specific automatic features have been added to the standard master/slave systems, and finally autonomous robots. Furthermore, the transporters, needed to position the manipulator where the task has to be performed, must also be considered (either a X-Y-Z crane system, or a more versatile mobile vehicle, for instance).

3.1 Dedicated machines

Operation aspects in the nuclear plants have lead to the design of specific remote tools, dedicated to a specific task. These tools can sometimes have a high degree of sophistication, and for instance include automatic features, similar to robotised equipments. A typical example is the fuel handling machine used in PWR reactors, or some steam generator maintenance systems. These machines are remotely controlled, and present some level of autonomy.

However, they are so specific to the task they have to perform, that any unforeseen circumstance lead to eventual man-entry or the use of other tools. Moreover, their installation requires usually also manual hands-on operation. Their application is therefore limited to very well defined repetitive tasks, appearing on a regular basis during the operation and the maintenance of a nuclear plant. Although they are an interesting application field for intelligent and sensor-based control, they lack the basic capabilities of human operators, w.r.t. flexibility, dexterity and failure rescuing, needed for any complex, and partially unpredictable operation.

3.2 Master/slave manipulators

Master/slave manipulators are used on a standard basis in hot laboratory cells. They operate through the walls of the shielded cells to perform dexterous tasks in line-of-sight applications. The master/slave arms are mechanically coupled with, for instance, steel tapes, which allow also forces to be transmitted. These systems remain unchallenged both in price and performance in many existing facilities. However these systems suffer from basic limitations, w.r.t. volume coverage, manual control necessitating direct sight, limited load capacity.

3.3 Advanced telemanipulators

In order to overcome the limitations of mechanical master/slave manipulators, new developments have been realised. Electric servo-manipulators were installed, allowing higher pay-load, larger distance between master and slave, and computer aid capabilities (33)(89)(90). Force control can be achieved by proper transducers on the slave arm, and the environmental perception is not limited to line-of-sight anymore. Video cameras, and sensors can inform the operator on the work space. The computer controlled manipulator can at the same time protect the machine from overload, or prevent it from entering forbidden space. It can extend the operator aid to choice of control modes, scaling movements and forces, performing subtasks in a complete autonomous way, etc (91). In fact, advanced teleoperators anticipate a more elaborated robotic mode, where generalised preprogrammed features, and automatic environmental perception, enhance the normal man-in-the -loop control, in order to improve both efficiency and reliability.

3.4 Telerobots

Industrial robots offer remote handling alternatives (34). They are generically very stiff, may have large lifting capacities, and are capable of high accuracy and repeatability. They benefit directly from the large amount of development effort invested in industrial automation, w.r.t. control, sensing and tools.

Their wide availability and low prices make them quite attractive for many nuclear applications. In nuclear applications, however, characterised by unstructured environment, automatic operation of such robots as applied in industrial production lines, is not feasible. The robot must be controlled using a remote man-machine interface, allowing mixed control mode (human operator/computer). Such a combination of robotics and teleoperation is often called telerobotics or telemanagement (96). Nevertheless, typical applications up to now have been limited to the kind of jobs similar to the ones appearing in manufacturing industries, and where contamination and radiation is low enough to allow hands-on maintenance. Applications concern for instance waste handling and inspection, where they provide a significant contribution to dose decrease. Developments are presently focused on improving the radiation resistance of such machine, in order to allow their use in a broader range of tasks, such as decontamination, dismantling or cell operation (27).

3.5 Automated gantry crane systems

One of the main problem of any advanced manipulator is to allow it to reach the required location to perform its specified task. In cell work with a very defined work space, the arms reach can be designed so as to have the fullest necessary coverage, and a fixed attachment on the wall is sufficient. However, in order to increase the cell dimensions, or to cover a more extensive work area, it is often very interesting to suspend the manipulator arms to some crane system, having one to three degrees of freedom (cartesian or cylindrical). Such a gantry can be controlled in the same way as the manipulator itself, and seen as supplementary joints of the cinematic system. Applications are however limited to well defined work areas, as it is generally the case in hot cells, waste handling department and fuel fabrication or reprocessing plants. Some work to be performed on the reactor vessel (top or interior) falls also in this category.

3.6 Long reach arms

In reactor plants however, such a disposition is not always feasible, as tasks must be performed in restricted areas where a gantry system cannot be envisaged. Moreover the complexity of the access path often require a very high level of dexterity to avoid obstacles. In these cases, an interesting alternative is to increase the number of degrees of freedom of the manipulator itself and to extend its range (so called snake manipulators) (35). This results in a high level of redundancy for normal operation, but this redundancy allows one to follow very complex paths in restricted geometries. The computer control of such systems however is quite demanding, both w.r.t. the cinematics itself than to the overall management and decision taking, as manual operation must be usually limited to the end-effector control only (39)(97).

3.7 Mobile platforms

In a broader type of applications, the manipulator must be transported through the plant, in an environment typical to human access. Doors, stairs must be coped with. These applications concern all types of tasks where the telemanipulator is meant to replace the usual human intervention (inspection, post accidental situations, etc). A mobile platform must be considered, and presently a considerable effort of research and development is devoted to this domain. The main focus has been put on the propulsion (tracked, legged, wheeled), communication (umbilical, wireless, in-board autonomy) and navigation (obstacle avoidance, obstacle recognition and coping with any route discontinuity) (23). Special versions of mobile platforms have been developed for particular environments, as for instance underwater robots (swimming devices or bottom crawlers) and in-pipe inspection machines (pipe crawlers).

4. Particular comparison between telemanipulators and telerobots

Some debate has arisen on the relative merits and future needed developments, between the traditional master/slave approach of the nuclear industry and the way followed by industrial robots (37). Actually these two approaches follow a very different historical path, but recent technological developments push them more and more to a common convergent point (15)(38).

Teleoperation comes from the concept of tools, allowing an operator to stay away from the actual work area (the blacksmith's tong). The master/slave manipulator is actually nothing more than a sophisticated tool. Their developments were closely related to the nuclear industry, because a direct necessity (no alternatives) to work remotely on very complex tasks appeared right at the beginning. Telemanipulators were seen as tools, prolonging the human operator, not replacing him. It is to be noted that the tasks to perform were also right at the beginning, complex, ill-defined a priori, unstructured, and that automation was not feasible at this early technological state (40).

On the other hand, robots were developed from the beginning as substitutes to man, for completely automatic functioning, but limited to well specified tasks, as they appear in fabrication processes. These tasks were simplified by adapting the environment to the robot, by structuring it in such a way that no high level perception and intelligence is needed (12)(34).

Advanced teleoperation tends now actually to include robotic features into the control, in order to

- substitute human control for specific tasks

- perform parallel tasks (controlling a camera, keeping a constant orientation) together with a human controlled action.
- check the human actions or modify them (obstacle avoidance, force and movement scaling, etc.)

An additional capability is in informing the operator on the state of the environment in the most efficient way, and to help him in taking decision (41)(42)(43)(44).

On the other hand, robot developments tends to automate more and more complex tasks, becoming less dependent on any well defined environment, making the machine capable to adapt itself to unforeseen situations, and to take decisions. As the state of the art in this direction does not allow up to now to consider total autonomy in completely unstructured environments, robot developments are forced to restrain their ambition and to introduce some kind of human control capabilities (take the difficult decisions, rescue after errors, etc) (27). This tends for complex problems, to rejoin the advanced teleoperation systems, and the word telerobotics has been suggested as a way to name this mixture of telemanipulator and robot concepts.

Present nuclear bilateral servomanipulators are not directly adaptable to robotic features. They suffer from the following drawbacks :

- poor accuracy in position and velocity
- closed control architecture not suitable directly for high level programming
- limited pay-load if high dexterity is needed. Heavy duty manipulators exist but lack force reflection, speed motions, accuracy and flexibility (degrees of freedom)
- decontaminability. Usual designs are open mechanics, necessitating gaiters protection.

One development direction is therefore to improve the present concept, to get a high payload, accurate, bilateral design with a high level of reliability and allowing full robotic control.

The second development line starts from industrial robots. These robots were not designed for work in nuclear environment, and therefore suffer from specific drawbacks :

- difficult remote maintainability
- poor radiation resistance
- difficult decontaminability
- no man-in-the-loop control, or no user-friendly man-machine interface

This second direction tends to improve existing industrial robots by alleviating these problems: replacing sensitive components by rad-hard alternatives, allowing easy maintainability by modularity, improving joint sealing and adapting man-machine interface with force feed-back and ergonomic design.

In comparing these two development directions, one must consider the type of application in mind. Adapted robots will get higher reliability and accuracy, and lower prices, as they start from proven technology, and large basis R&D background. They will beneficiate from all advances in the domain of intelligent control achieved for the computer aided manufacturing.
Advanced telemanipulators will have a much better payload to weight ratio and will offer easier transport, access and deployment capabilities. They will be better suited for constrained spaces and mobile transporters, and easier to be adapted to special situations, such as underwater work for instance.

5. Teleoperated tasks in decommissioning work

Decommissioning work of nuclear facilities involves a broad range of activities requiring the ability to dismantle major equipments in the presence of radiation field and contamination (47)(48)(85). Many of these tasks are generic to most nuclear applications, in particular the inspection tasks and the waste handling. Others are more typical, and actually are sometimes unique in the range of applications for robotic machines, tackled up to now. They mostly concern the dismounting, cutting and general size reduction operations, as well as surface chemical decontamination or mechanical removal. The different tasks can be listed as follows :

5.1 Survey

Components to be removed must be surveyed to assess the degree to which they are contaminated. These survey must also evaluate radiation fields in the vicinity of these components as well as contamination levels to potential work areas. Moreover, the mechanical state of the equipment must be evaluated or measured in order to assess the dismantling strategy.

Teleoperation can be considered here in the same way as for inspection tasks in nuclear power plant. The requirements are similar, w.r.t. access, sensing needs, etc. The radiation levels are usually low, and teleoperation is only an alternative to hands-on operation. However, for specific tasks related to highly radioactive parts (fuel, reactor vessel, debris), complex tele-operated sensing systems must be considered (45).

As for nuclear inspection, two types of remote equipment can be envisaged here: the custom-designed teleoperator adapted to a given particular geometrical environment, or an automatic guided vehicle capable to survey a larger range of environments. However, as these machines appear here usually to be mere alternatives to hands-on solutions, developments of specific machines will not be justified, but their use will depend usually on the availability of existing ones developed previously during the plant operation.

5.2 Work area preparation

Any actual decontamination or dismantling operation must be preceded by a preparation of the work site. Protective barriers must be placed (shieldings, hermetic seals, etc), and needed tools must be installed. Classical operation or maintenance tasks must be performed on the equipment to put it in an adequate and safe condition (isolation, emptying, removal of auxiliaries, etc). Finally, the actual teleoperated machine must be installed and connected, with the related sensing and viewing systems.

These operations are representative of many activities in the nuclear industry and do not set special requirements in the case of decommissioning context. It should be however stressed that here the teleoperator is installed usually for a shorter period, for one specific job. Therefore the installation covers a proportionally higher part of the whole mission time and should be optimised, both in downtime and dose exposure needs (46)(49). In this respect, the teleoperated equipment ideally should be flexible enough to be usable not only for the proper job to perform, but also for the related preparation work.

5.3 Decontamination

Components or parts of the infrastructure must be decontaminated either by a surface cleaning process, or by mechanically removing the contaminated layer. These processes generate a large amount of dispersed radioactivity in the cleaning liquid or as air dust. This activity is collected through filters or other antidispersant equipment, that must in turn be handled to be replaced and transported away as waste (50).

Teleoperated surface decontamination involves requirements similar to other survey activities, w.r.t. access capabilities and positional accuracy (29). In some cases (for instance microwave or explosive concrete stripping), debris collection is a related task sometimes more demanding in terms of advanced teleoperation (gripping strategy, debris recognition in unstructured environment).

Usually, custom-designed teleoperated equipment will have to be used. But, in some instances, where large wall areas must be cleaned (hot cell decontamination, pool scrubbing, concrete wall stripping), nuclear versions of industrial robots can be interesting for their lower cost and higher flexibility (27)(34).

5.4 Dismantling

Dismantling tasks have as objective the size reduction of large equipments in order to ease their transport and their storage as waste. In limited cases, this can be achieved by normal dismounting of mechanical connections. However, in most situations, there is no alternative to the cutting of oversize components into smaller pieces. Such cutting operations can be performed by a great variety of techniques, involving mechanical contact (sawing, grinding, nibbling, hammering) (52)(55) or not (plasma arc, water jet, electro-erosion, laser, etc) (51)(56)(57). These operations are sometimes to be performed under water.

These types of tasks present the highest specificity w.r.t. classical use of telemanipulators and robots. Actually, the dismantling process generates itself an unpredictable environment, inherently increasing therefore with time the level of complexity and uncertainty. Particular attention must be put here on equipment holders and debris collection. In some cases, two (or more) tele-operated machines must be involved for this purpose, and cooperative control of these machines must be insured (54)(59)(63)(94).

Another specific problem is the level of force reflections that can be generated when contact cutting is performed. These important efforts can cause jamming problems of the tools, trajectory instability and even collision damages. In some cases, force control is even necessary to avoid failure.
Comparing with similar tasks appearing in other industrial activities, one can also stress on the other hands relaxing requirements, as cutting speed or cut accuracy characteristics is less critical. But cutting parameters must nevertheless be closely controlled as they influence the secondary waste production.
Remote equipment for this purpose will usually be custom-made advanced teleoperators, especially when specific environmental conditions are to be met (underwater work, high cutting forces, high radiation field, restricted access) (53)(58). For dismantling of lighter structures, such as glove boxes for instance, nuclear versions of industrial robots can offer interesting alternatives.

5.5 Waste packaging

This includes moving radioactive and contaminated items to the packaging area, as well as the actual packaging itself. Items can include components and pieces removed from the plant, decontamination wastes, tools, etc. The waste is introduced in cans and containers, eventually in shielded drums. These packages must be closed (bolted, sealed, etc) and the cask surfaces must be checked for external contamination, before to be released for storage. Transfer systems must be used to transport the waste to the storage facility (60).

Teleoperation finds here more conventional task definitions, corresponding to similar automated warehouse applications (27). As the level of radiation is also usually moderate, waste handling offers the most adequate environment for industrial robots installation.

6. Major issues and requirements

The tasks that have been described above set a series of special requirements on teleoperated equipments to be used for decommissioning activities. Some of these requirements are common to other nuclear applications. However, due to the particular nature of the dismantling tasks, new specifications appear with respect to the man-machine interface and equipment autonomy, to the environmental tolerance and reliability, as well as to the radio-protection optimisation.

6.1 Man-machine interface

An advanced teleoperated equipment is actually a combination of four elements: the slave arm installed in the work area, a computerised control system, the master arm with auxiliary informative systems, and the operator. The control mode combines automatic features with manual operation. The man-machine interface is the fundamental link of the system allowing maximum coherence between these two levels of control. The interface must provide three functions depending on the task to be performed:

- to inform the operator on all data needed to get a better perception of the state of the slave part of the system and its environment (real video viewing, simulated scenes, force feedback, synthetic/graphical presentation of key measured data)
- to substitute automatic control for manual one in some physical actions (automation of repetitive tasks, parallel control of auxiliary equipments as cameras, collision avoidance features, trajectory and orientation following)
- to aid the operator in decision making.

In some dismantling tasks, the man-machine interface will be far removed from the actual work zone, without direct line-of-sight possible, and with difficult viewing capabilities (restricted geometries, poor lighting conditions under water). The importance of a reliable and user-friendly control station is therefore here especially critical.

6.1.1 Ergonomy of master control station

The design of the master device is essential in allowing for easy operation. Its ergonomy must be such as to allow for unconscious perception of the real geometrical position of the slave equipment. Movement should be given in a coherent way (61). But at the same time, the master device must be simple.

Trying to duplicate on the same device all the degrees of freedom of the slave arm lead usually to excessive complexity, and a lack of global control capability (65)(66). In some cases, the presence of two operators is necessary to control different parts of the device (for instance the end-effector and the viewing system). For complex tasks, however, proper training of the operator is always very important and must be insured, on real-size partial mock-ups (67).

6.1.2 Force feedback capabilities

In dismantling tasks, high forces are sometimes generated by some cutting tools. These forces can cause damage to the surroundings, to the tool itself, or degrade the cutting parameters. A proper control of these forces is essential. The man-machine interface must therefore provide the necessary force feedback (eventually with a chosen scaling factor and with offset compensation) (33). The forces must be measured on the slave arm (strain-gages, instrumented wrist, deforming flexible structures) and reproduced to the master device. The feedback capability can also be used to provide artificial proximity sensing in case of forbidden zones.

6.1.3 Assistance to operator's decision

Due to poor remote environmental perception, the operator experiences sometimes difficulties in taking decisions, w.r.t. tasks sequences, proper timing or go clear evaluation. The man-machine interface should allow for a level of computer-aided-decision (40). But taking into account that, for the complex tasks involved in many dismantling work, the operator must keep the final decision role and that decision aids must be limited to suggestions of actions or double checks before actions.

6.1.4 Human errors avoidance

Reliability and accident analysis have shown that the main risk factor in teleoperated equipment lies in human errors, caused by uncompleted perception of the environment and the state of the tasks performed by the slave or by inattention due to stress and fatigue (64). In the case of most dismantling operation, consequences of human errors can be far reaching, costly both in time and eventual radiation dose. Jamming of tools, lack of good debris management, etc, can generate complex unforeseen situations, not easily handled by the available remote equipments. The man-machine interface should in some extend screen out the manual orders to filter out any doubtful, forbidden or risky step. Typical examples are concerning collision avoidance or allowed movement orientation (in transporting liquid waste, in cutting tool orientations, etc). The sequence of actions should also be monitored by a clearing process, checking before each task is launched if the state of the equipment and the environment is in the right condition.

6.2 Environmental perception

In many dismantling tasks, the environment is poorly known, and the conditions of perception are difficult. The operation reliability depends thus on all the data sensed around the work area and the way they are reproduced for the operator.

6.2.1 Remote vision

Remote operation rely heavily on vision perception through video images. For particular dismantling operation, especially under water in constrained space (poor quality images), or in highly clustered environment (lack of depth perception), it is often difficult for the operator to get a right perception of the relative position of all objects (70). A certain number of improvement are therefore of primary importance:

- stereo capability
- enhanced images (contrast correction, noise reduction)
- multiple cameras, with at least general view and detailed view.
- multiple degrees of freedom for the camera positioning, including zoom capabilities. Actually, the cameras require usually a teleoperated arm themselves, eventually controlled in parallel by a dedicated operator and computer.

6.2.2 Computer image simulation

The use of real images to get a perception of the environment has several limitations, especially under poor lighting conditions, and occlusion problems. In the case of a known or partially known environment, a useful source of information can be obtained by a graphical computer simulation of the slave device in its environment. This allows the operator to get a very clear image to follow, to virtually zoom to any detail without physical constraints (camera support, occlusions, etc). It can be used to on-line follow a given operation, but also to simulate in advance any particularly tricky task (62)(69)(72). Care must however be kept in the level of confidence of such techniques in the case of dismantling work, where the environment is subject to progressive degradation w.r.t. the initial model.
A possible way to improve the level of confidence is to superpose real and synthetic images on the same screen, or to juxtapose them on adjacent ones. The operator can then easily assess the model validity at regular intervals.

6.2.3 Model based control

The same caveat should be expressed on any other model based control aid, as for instance model based object recognition, collision avoidance, trajectory following, etc. As such techniques are feasible for inspection, decontamination or waste handling work, they should, at the present state of technology, find limited application in real dismantling tasks.

6.2.4 Sensor based control

In many cases, a model can be used to aid the environmental perception, as long as this model can be maintained, updated, corrected using physical data measured on the spot. Position sensors can be used and their results compared to the model (68)(71)(95). In some cases, unknown obstacles can be roughly reconstructed using these data and added to the model. Confirmation by real images can be decided at operator level. Sensors data can be reported directly to the operator, eventually in a tangible way (virtual repulsion force in case of obstacle proximity, audio signals, simulated attitude display for the end-effector, etc). They form naturally an essential part of any automatic control feature.

In many decommissioning tasks, the availability of two types of sensors is needed: non-contact proximity sensors and force (and momentum) sensors. These sensors should be located mainly at the end-effector level, but also at particular joints of the manipulator.

6.3 Technological challenges

The tasks to be performed, and the degree of automation wanted, require a certain number of advanced features to be present on the teleoperated machine. In typical decommissioning tasks, environmental constraints will however present particular challenges w.r.t. present state technology. The constraints are mainly related to the radiation field, the underwater environment, the data communication over long distances and the high level of reliability required.

6.3.1 Radiation hardening

One of the most demanding challenge to advanced nuclear teleoperation is the tolerance of sophisticated equipment to the radiation field (70). For decommissioning works, only gamma radiation will have to be taken into account. No neutrons are present, and alpha and beta radiation can be easily shielded out. In most high dose cases, this gamma field lies mainly around 1 MeV, as Co isotopes form the highest source term. The energy dispersion is much more pronounced in applications where the dose rate is lower. It is therefore usual, and justified, to analyse the tolerance of sensitive equipment to Co-60 gamma rays.
Both total dose and dose rate is important to consider, as radiation tolerance is a non-linear phenomena, where degradation and annealing mechanisms are concurrent.
It is difficult to set quantified values on doses to be expected in decommissioning tasks, as they are very dependent on the type of work, the vicinity of highly radioactive components (near-core vessel parts), the degree of decontamination achieved, and the mission times of the equipment in the work area.

However, one can roughly consider three categories of environment:

- Highly radioactive, when components around the reactor core must be removed and cut, or when fuel elements must be handled. Typical dose rates can reach 1 to 100 krad/h, in close vicinity of the components, and total dose can be as high as 1 to 100 Mrad, considering mission times of up to 1000 hours.
- Medium doses, when most components of the primary loop of a reactor are handled, or when decontamination waste is considered. Some glove box dismantling work can also fall in this category. Here, the doses are high enough to prevent man approach, but they are limited to dose rates of less than 1 krad/h, and total dose is usually kept below 100 krad.
- Low doses are typical of all applications where manual operation is possible, but where remote manipulation is considered as an improvement in radioprotection, efficiency and speed. These environments can present high contamination risks, but the requirements concerning radiation tolerance are mild, as dose rates are typically less than 1 rad/h, and total doses usual limited to 1 krad at the most.

Classical material and equipment design can be used mostly for any task related to the third category, as the second one will require screening and radiation hardened version of sensitive devices, such as electronics, cameras and sensors. These equipments are however in most cases commercially available. For the first category, only a limited sensing and electronics will be possible (special design, sometimes custom-made developments necessary), and alternative solutions must be envisaged (73)(75). The positioning of cameras and sensors at a given distance from the actual radioactive source, at least during idle periods can be a solution. A trend is now appearing to allow rather for easy replacement of sensitive parts, with throw away equipments being seen as a cost-effective alternative to rad-hard designs (74).

6.3.2 Underwater tasks

Many dismantling tasks to be performed on highly radioactive components, as for instance on the reactor vessel and on all the core internal supporting structures, are conducted under water. This allows both for easier contamination spread minimisation and for biological shielding of the remotely located operators. However, under water tasks are demanding supplementary care on any teleoperated equipment (55)(58)(71). Water leak tightness up to several meters require specific joints, casing, instrumentation penetrations. Buoyancy forces must be compensated, and lighting conditions are to be adapted (higher light absorption, change in refraction index). Moreover, in some cases, the dismantling process itself is generating water fouling in such a way to impair greatly any viewing.

Particular attention must also be put on the drive mechanisms of the teleoperated machines. Electrical drives require proper tightness to avoid electrical hazard, and hydraulic drives can potentially contaminate the surrounding water in case of leakage. Proper choice of drive liquid is to be made. usually water based compounds are preferred.
One particular aspect related to under water operation is that the biological shielding cannot be used to support the equipment and the adjacent umbilical, as it is the case for through wall operation. Both the equipment and the power and signal cables must be supported over long distances. Umbilical management is therefore here usually more difficult, especially when complex geometries and constrained space are present.

6.3.3 Data communication

The problem of the management of the umbilical appear in fact in any teleoperated equipment having a somewhat large work envelope. In particular cases, the cables can be made part of the mechanical structure, but in many applications they form an adjacent bundle requiring particular supporting and handling care. It is therefore of primary importance to reduce the number of cables as much as possible. Two ways can be followed:

- multiplexing the signals by installing power distribution manifolds on the slave arm, multiplexing sensor signals (analogue or digital multiplexers) and multiplexing video signals over large band communication links (eventually with optical fibres). In highly radioactive environment, such solutions will be challenging, as practically all performing multiplexing techniques assume the presence of sophisticated electronics in the radiation field. Present state of the art offers however signal multiplexing solutions up to a total dose rate of 1 Mrad, both electrical or optical, using special rad-hard components. The multiplexing of video signals is on the other hand not yet feasible over a large scale in these environments. It is to be noted that this aspect greatly beneficiate from the rapid developments made in telecommunications for space applications.
- using wireless communication. However, this solution is usually quite difficult to implement in steel reinforced concrete building, with sometimes steel plated walls, and high level electrical noise background. Only particular applications, similar to the ones appearing in automated factory can be envisaged (waste handling, some inspection tasks).

6.3.4 Reliability and safety

An advanced teleoperated equipment used for decommissioning work represents a complex functional system with particular potential failure risks, due to
- the non-structured environment, usually not designed for such an equipment
- the environmental influence on the functionality of the components, mainly due to radiation levels
- the mixed mode control involving two levels of hierarchy: man and computer.
- the process involved in some cases, as for instance for the cutting tasks, and representing hazards, w.r.t. fire, contamination or mechanical damage.

On the other hand, any failure can have important consequences, as the work space cannot usually be entered without special precautions, if any. This would always represent high costs, long down-times and supplementary radiation doses on personnel.
The requirements on reliability are therefore very stringent, and a reliability model should be constructed for each important task to be performed. This would allow one:
- to quantify the risk probability
- to list the different failure modes
- to envisage the different recovery paths available in case of failure
- to improve the critical paths, w.r.t. their failure risk (higher reliability) or consequences estimation (mitigation techniques and back-up solutions).

Particular attention should be put in dismantling works on two risks, specific to these applications:
- the fire hazard, as cutting involves local high temperatures and sparks
- the handling of heavy (sometimes not well defined geometrically) components and debris.

6.4 Radiological optimisation

Implementation of advanced teleoperation is mostly thought in a context of radioprotection optimisation (77). Except for specific situation where no other alternatives exist due to high radiation dose rates or constrained space, teleoperation is evaluated w.r.t. human hands-on operation (79). Although cost, time and reliability are also critical parameters in this evaluation, radioprotection optimisation encloses them in their most part, as operators preparation, work shifts, radioprotection control are directly linked to high costs and delays.
One should however not be mislead by a naive image, where remote operation is the overall solution to any ALARA based approach. Present state-of-the-art machines are not versatile and reliable enough to exclude any human intervention during their application.

This is particularly true for many of the decommissioning tasks, as they are usually far different from usual industrial experience in remote operation. In particular, the machines must be installed. They must be rescued in case of failure or emergency. They must be decontaminated prior to any maintenance work or transportation. And finally they must be repaired, adapted, upgraded to changing situations and needs. All these activities mean human interventions, sometimes inside the radioactive work space.
Therefore, a thoroughly conducted work analysis has always to be performed, taking into account all events, both planned for the operators and the equipment, and unforeseen (with the best probability estimates) (82).

6.4.1 Manual installation versus mobile platform

An important part of the human intervention around any teleoperated intervention is taken by the work space preparation and machine installation. Important radiation doses can be involved with these works, especially in highly radioactive and contaminated locations. In this respect, mobility of the equipment to get in place and install itself is a key feature for many tasks. Gantry systems extending outside the dangerous zone can bring interesting solutions in particular situations. Completely autonomous mobile platforms (76) adapted to the geometrical environment are usually more demanding w.r.t. to the present technological capabilities, but important R&D efforts are devoted to this critical issue. Without going however as far as automatic work space preparation and installation, the equipment design and work schedule should take these tasks seriously into account, and optimise them both in time needed and simplicity of execution (78)(80).

6.4.2 Rescuing capabilities

Teleoperated equipments are complex systems, in their mechanical and electrical parts, in their control algorithms involving different levels of hierarchy, in their ill-defined environment and task specification. Although careful design can enhance their reliability to high values, failure cannot be excluded. This can mean the failure of an equipment component, the jamming w.r.t. the surrounding objects, a control failure, or a human error. Consequences of such an event can be sometimes far reaching, and therefore attention must be put on two aspects:

- early failure diagnostic, using adapted sensors or by providing the operator with a continuous quick global view of the equipment functioning.
- predefined back-up procedures to retreat to a safe position (e.g. with partial drive capabilities, with supplementary operator driven tools, etc.)

A serious failure event tree analysis is a prerequisite of primary importance to provide a comprehensive set of these back-up procedures.

6.4.3 Decontaminability

Any repair or scheduled maintenance operation on equipment used in highly contaminated environment, will normally be preceded by a decontamination phase. This decontamination can be a source of human dose build-up and contamination risk (81). The equipment design must therefore be adapted to rapid and easy decontamination (flat surfaces will non-porous finishing, sealed joints, etc.). This easy decontaminability will have a determinant impact on the radioprotection optimisation throughout the whole life cycle of the teleoperated machine.

6.4.4 Maintainability

Depending on the type of application and physical environment, some maintenance or repair work has to be performed on the spot. The whole equipment cannot be easily removed out of the work zone, or the delay involved in such removal is inefficient. In such circumstances, the maintenance work must be optimised in down time (i.e. in radiation dose exposition):

- easy replacement procedures of failure-risk components (83),
- modular construction of the equipment allowing for easy removal of one element only (for instance one link of an articulated system) (27)(76).

6.4.5 Training

As it is the case for any operation under hazardous environment, proper training of any human intervention is of primary importance. This training must be performed on realistic mock-ups representing the major features of the real work space, including their modifications with time (67). This is particularly true for dismantling operations, where the environment can change drastically during the process, and where a periodical updating of the mock-up, and of the training, will be needed.

7. CEN/SCK involvement in the domain of dismantling and teleoperation

As many other nuclear instalments established in the early days of the nuclear industry, the CEN/SCK is presently confronted to some shut-down installations, that must be in one way or the other set at least in a low-surveillance stable state, or even dismantled. The decision to really dismantle big installations as reactors for instance is difficult, as divergent arguments can be expressed, for immediate dismantling, postponed decision, or even to consider the reactor containment as a safer storage alternative. As a research centre, however, specific research projects are under way related to the problematics of decommissioning in a broad range of domains (decontamination, waste characterisation and handling, radioprotection optimisation, cutting techniques, teleoperation).

Only pertinent projects to the subject of this paper are summarised hereafter.

7.1 BR3 dismantling pilot project

The European Community Decommissioning Programme is presently funding four pilot real size dismantling projects, one of which concerns the BR3 reactor in Mol. The BR3 is a 10 MWe PWR reactor that was operated from 1962 to 1987, mainly as training installation for reactor personnel and as test reactor for advanced fuel. Starting in 1989, a dismantling project is now under way, and it places BR3 as the first European PWR to be dismantled. During the current phase of this project, three basic tasks are being performed:

- the decontamination of the primary cooling loop (terminated in May 1991),
- the cutting of the thermal shield (mid 1991), a large stainless steel cylinder (6m high, internal diameter 1.3m, thickness 75 mm) that will be cut by mechanical sawing (horizontal cuts) and plasma arc (segmentation of cut rings). Siemens-KWU and Framatome are associated contractors for this part.
- the disassembly and size reduction cutting of all remaining internal components of the vessel (core supporting structures) (starting in 1992).

The cutting of the thermal shield involves a dedicated remote operated machine allowing for the mechanical cutting, the extraction of the cut parts and the collection of secondary waste. The rings are then segmented by plasma arc in a specially designed isolation chamber, using a three degrees of freedom manipulator arm. All these operations are performed under water.
The disassembly and cutting of the other internals will involve remote operated machines (both custom made tools and computer aided flexible arms) using plasma arc, electro-erosion and mechanical techniques. The complex geometry of some components will offer a particular challenge to the underwater remote operation (84).

7.2 Telerobots for decontamination and cutting operations

The CEN/SCK together with AEA-Technology in Harwell is presently involved in R&D work on the use of existing industrial robots for monitoring purposes, decontamination of flat surfaces and dismantling of glove boxes. The project called TMDSRS (for Telerobotic Monitoring, Decontamination and Size Reduction Systems) is funded by the European Community in the frame of the Decommissioning Programme and will run up to the end of 1993. The project proceeds in two distinct stages involving first the continued development of the NEATER Robot (27) and its application in cold trials, and secondly an active trial in one of the three types of activity.

The cold tests consist in
- contamination mapping and decontamination using an electro-polishing head unit developed at Harwell.
- monitoring nominally clean surfaces of equipment prior to clearance as uncontaminated.
- size reduction of glove boxes in work cells.

The CEN/SCK is mainly involved in the sensor based control aspects and the radiation tolerance analysis and tests of the end effector parts.

7.3 Vision and sensor systems

Under the frame of the European Community TELEMAN programme, CEN/SCK is coordinating one of the 17 on-going projects. It is called VISYS for Vision and Sensing Systems and runs for two year up to the end of 1992. The other partners of the projects are CRIF in Brussels, KfK in Karlsruhe, AEA in Harwell and the University of Oxford. Its global objective is to develop all needed components of an integrated and performing environmental perception system for semi-autonomous teleoperation. It constitutes one element of a package of four linked proposals, called CIRCUIT, for the development ultimately of an intelligent telerobotic gantry/-manipulator system. Detailed work is being performed on
- vision in poor lighting conditions (flame viewing, shadowing)
- position detection and collision avoidances sensors
- 3D object localisation and recognition
- graphical scene representation
- data fusion for global position determination

The CEN/SCK is mainly involved in proximity sensors requirement analysis, and related developments.

7.4 Radiation tolerance testing

Another TELEMAN project, called ENTOREL (for Environmental Tolerance, Reliability and Safety), is aimed at the study of the radiation tolerance of robotic components used in nuclear tele-operation. It also addresses the issues of reliability analysis and safety assessment. The project has been chosen as a backbone of the TELEMAN programme for all problems related to reliability and safety, extending therefore (up to the end of 1994) into the second phase of the programme devoted to the construction of prototype machines. ENTOREL is coordinated by RISOE in Denmark, and groups besides the CEN/SCK, also Interatom (Germany) and AEA in Harwell. Two goals are particularly addressed:
- Radiation tolerance
    * Constitution of a radiation resistance data base
    * Testing campaigns for specific components (electronics, sensors, multiplexing systems, vision systems, driving mechanisms)

- Reliability and safety
  * Modelisation of failure modes due to radiation induced degradation
  * Study of failure correction strategies and man machine interface optimisation

The CEN/SCK contribution consists mainly in representative testing of sensors and communication equipments in the gamma facilities connected to the BR2 reactor.

7.5 Remote maintenance for fusion reactors

CEN/SCK is also involved for the last five years in R&D work related to the remote maintenance of the future NET fusion reactor. Together with three other partners (KfK in Karlsruhe, CEA in Fontenay-aux-Roses and ENEA in Casaccia) the in-vessel handling unit is studied in a pre-design phase. The CEN/SCK is mainly involved in radiation tolerance testing at the high radiation levels (3 Mrad/h) and temperatures (150°) foreseen in the vessel during shut down maintenance. A series of gamma irradiation is presently under way concerning :

- position sensors and multiplexing electronics
- camera electronics and optics
- drive components
- connection material, including optical fibres

The tests are foreseen to be continued at least up to the end of 1993.

8. References

(1) Robotics and Nuclear Power, Report by the Technology Transfer Robotics task Team US DOE Document NE-0065, Dec. 1986.

(2) L. Poulter, et al., In-service Inspection Techniques for PWR Steam Generator Feedwater and Pressuriser Nozzles, Proc. International Conference on Remote Techniques for Inspection and Refurbishing of Nuclear Plants, Stratford/Avon, Nov. 28 - Dec. 1, 1988.

(3) J. Seguy, Robotics in French PWR Nuclear Power Plants, Conference on Remote Techniques for Inspection and Refurbishment of Nuclear Plant, Stratford-upon-Avon, Nov. 28 Dec. 1, 1988.

(4) K. Sawaragi, et al., An Underwater Vehicle for Ultrasonic Vessel Inspection, Nuclear Engineering International, Dec. 1990, pp. 40-41.

(5) P. Krausser, Methods and Equipment of Remote Controlled Repair for Steam Generator Tubing, Conference on Remote Techniques for Inspection and Refurbishment of Nuclear Plant, Stratford-upon-Avon, Nov. 28 Dec. 1, 1988.

(6) P. Smith, The Warrior Welding Manipulator, Second International Seminar on Robotics in the Nuclear Industry, Lausanne, Aug. 24-25, 1987.

(7) M. Obama, et al., A Locomotive Inspection Robot for Turbine Building Interior Inspection in NPP, International Conference on Advanced Robotics, Tokyo, Sept. 9-10, 1985.
(8) D. Friend, et al., An Overview of the Use of Mobile Inspection Vehicles within the CEGB, Conference on Remote Techniques for Inspection and Refurbishment of Nuclear Plant, Stratford-upon-Avon, Nov. 28 Dec. 1, 1988.
(9) J. Connor, The Development of a Steam Generator Inspection Robot, Conference on Remote Technology for the Nuclear Industry, London, Dec. 10-11, 1990.
(10) A. Collie, A Climbing Vehicle for Nuclear Reactor Inspection, Conference on Remote Technology for the Nuclear Industry, London, Dec. 10-11, 1990.
(11) F. Champigny, et al., Cuve principale de Creys-Malville, Bilan de l'inspection en service réalisé par le MIR, RGN, No. 2, Mars 1990, pp. 116-119.
(12) J. Fisher, Applying Robots in Nuclear Applications, Proc. Robot-9 Conference, Detroit, June 1985.
(13) H. Okano, Intelligent Robots for Nuclear Plants, Second International Seminar on Robotics in the Nuclear Industry, Lausanne, Aug. 24-25, 1987.
(14) W. Pauli, The Surveyor Mobile Robot Vehicle, Second International Seminar on Robotics in the Nuclear Industry, Lausanne, Aug. 24-25, 1987.
(15) P. Jezequel, Téléopération et robotique dans les centrales nucléaires, perspectives et réalités, ORIA-89 Conference on Remote Control in Hostile Environment, Marseille, Dec. 11-12, 1989.
(16) S. Hosaka, et al., Teleoperation of Robot for Maintenance and Inspection in the Containment Vessel of NPP, International Conference on Advanced Robotics, Tokyo, Sept. 9-10, 1985.
(17) T. Collis-Smith, The Application and Development of Robotics to Some Remote Inspection and Maintenance Tasks, taken from Britain's AGR's., Conference on Remote Technology for the Nuclear Industry, London, Dec. 10-11, 1990.
(18) D. James, Remote Replacement of Warwick Links at Oldbury NPP, Conference on Remote Technology for the Nuclear Industry, London, Dec. 10-11, 1990.
(19) M. Feldman, Design Concepts and Advanced Manipulator Development for Nuclear Fuel Cycle Facilities, First Seminar on Nuclear Robotics, 8th SMIRT Conference, Brussels, Aug. 26-27, 1985.
(20) T. Ohmichi, et al., Development of the Multi-function Robot for the Containment vessel of the NPP, International Conference on Advanced Robotics, Tokyo, Sept. 9-10, 1985.
(21) J. Draulans, et al., BelgoNucléaire Automates Mixed Oxide Manufacture, Nuclear Engineering International, March 1985, pp.,31-33.

(22) W. Issel, et al., Application of Large Cell Remote Handling Techniques in Nuclear Plants, Kertechnik, vol. 54, 1989, pp. 238-243.
(23) H. Meieran, Mobile Robots Continue to March, Crawl, Roll, Walk, Slither, Climb and Swim into Action, Nuclear Engineering International, Jan. 1991, pp. 26-27.
(24) E. Baumgaertner, et al., On the Possibility of Remote Monitoring of a Reprocessing Plant, Kerntechnik, vol. 55, No. 3, 1990, pp. 138-146.
(25) J. Berger, et al., The Use of Robotics for Radioactive Waste Shipping and Receiving, Sandia report SAND-86-1088C, Dec. 1986.
(26) R. Delrue, et al., Expertise in Automated Handling in Nuclear Environment, Second International Seminar on Robotics in the Nuclear Industry, Lausanne, Aug. 24-25, 1987.
(27) E. Abel, et al., A NEATER Advance in Active Handling Technology, Nuclear Engineering International, 1990, pp. 26-32.
(28) J. Goodman, Remote Sampling to Remove Metal Samples, Conference on Remote Techniques for Inspection and Refurbishment of Nuclear Plant, Stratford-upon-Avon, Nov. 28 Dec. 1, 1988.
(29) H. Meieran, Clean up Acts Conducted by Mobile Robots in Nuclear Facilities, Conference on Remote Techniques for Inspection and Refurbishment of Nuclear Plant, Stratford-upon-Avon, Nov. 28 Dec. 1, 1988.
(30) D. Merchant, et al., Post-accident Recovery Operations at TMI-2, Workshop on Mobile Teleoperators for Radiological Emergency Response and Recovery, Dallas, June 24-25, 1985.
(31) F. Schwarz, et al., TMI-2 Clean-up Techniques Applicable to Decommissioning, Proceedings of the International Decommissioning Symposium, Pittsburgh, Oct. 4-8, 1987.
(32) F. Schwarz, et al., TMI-2 Experience with Remote Operations Technology, Proceedings of the International Decommissioning Symposium, Pittsburgh, Oct. 4-8, 1987.
(33) G. Clément, et al., A New Manipulator Under Development in France, International Conference on Decommissioning of Major Radioactive Facilities, London, Oct. 11-12, 1988.
(34) E. Abel, Robotic Devices for Future Nuclear Plants?, UKAEA Report AERE-R-11822, May 1985.
(35) J. Holt, et al., Computer Control of a Multi-link Manipulator, Conference on Remote Techniques for Inspection and Refurbishment of Nuclear Plant, Stratford-upon-Avon, Nov. 28 Dec. 1, 1988.
(36) IAEA, The Clean-up of Large Areas Contaminated as a Result of a Nuclear Accident, Technical Reports series No. 300, Vienna, 1989.
(37) I. Salter, D. McCabe, Nuclear Industry Requirement for Robots, Vol. 1 and 2, Taylor Hitec Ltd, CEC Contract ECI-998-B7210-83UK, May 1985.

(38) H. Martin, et al., Joining Teleoperation with Robotics for Advanced Manipulation in Hostile Environments, Proc. Robot 8 Conference, Detroit, 1984.
(39) T. Tsuchihashi, et al., A Computer Aided Manipulation System for Multijoint Inspection Robot, International Conference on Advanced Robotics, Tokyo, Sept. 9-10, 1985.
(40) J. Vertut, P. Coiffet, Computer Aided Teleoperator Systems, a Major Step to Intelligent Manipulation and Locomotion, International Conference on Advanced Robotics, Tokyo, Sept. 9-10, 1985.
(41) W. Uttal, Teleoperators, Scientific American, Dec. 1989, pp. 124-129.
(42) J. Benner, et al., Towards More Automation for Remote Maintenance, IAEA-TECDOC-495, 1989, pp. 249-260.
(43) D. Grisham, et al., Application of Teleoperator Expertise to Robotics, Los Alamos Report LA-UR-85-4066, Dec. 1986.
(44) R. Lumia, et al., Teleoperation and Autonomy for Space Robotics, Robotics, vol. 4, 1988, pp. 27-33.
(45) W. Hamel, Autonomous Robots for Hazardous and Unstructured Environments, Oak Ridge National Laboratory Report CONF-860487-1, 1987.
(46) D. Ashcroft, et al., Engineering Design of the Windscale Advanced Gas Cooled Reactor Decommissioning Machine and Robotic Manipulator, International Conference on Decommissioning of Major Radioactive Facilities, London, Oct. 11-12, 1988.
(47) L. Da Costa, et al., Systems for Remotely Controlled Decommissioning Operations, Graham and Trotman, London, 1990.
(48) I. Munro, Telemanipulation and Telerobotics Applied to Decommissioning of Nuclear Facilities, Conference on Remote Technology for the Nuclear Industry, London, Dec. 10-11, 1990.
(49) E. Jones, Remote Diversion of a Highly Active Process Pipeline, Conference on Remote Technology for the Nuclear Industry, London, Dec. 10-11, 1990.
(50) J. Devore, Remote Methods for Decontamination and Decommissioning Operations, Oak Ridge Report CONF-860203-5, Dec. 1986.
(51) M. Contre, Robot Laser dans l'industrie nucléaire, International Conference on Laser Materials Processing Science and Applications, Osaka, May 21-23, 1987.
(52) S. Yanagihara, et al., Remote Cutting Systems for Dismantlement of Steel Structure in the Japan Power Demonstration Reactor, International Conference on Decommissioning of Major Radioactive Facilities, London, Oct. 11-12, 1988.
(53) G. Bernard, Machine de démantèlement Atena, ORIA-89 Conference on Remote Control in Hostile Environment, Marseille, Dec. 11-12, 1989.

(54) G. André, et al., Generalised End Effector Control in a Computer Aided Teleoperation System with Application to Motion Coordination of a Manipulator Arm on an Oscillating Carrier, International Conference on Advanced Robotics, Tokyo, Sept. 9-10, 1985.
(55) P. Antoine, et al., Underwater Dismantling of a Secondary Source Assembly by Teleoperation, Conference on Remote Techniques for Inspection and Refurbishment of Nuclear Plant, Stratford-upon-Avon, Nov. 28 Dec. 1, 1988.
(56) B. Migliorati, et al., Investigation of Laser Cutting Applications in Decommissioning, in Decommissioning of Nuclear Installations, Ed. K. Pflugrad, Elsevier, London, 1990, pp. 181-190.
(57) R. Léautier, et al., Development of a Prototype System for Remote Underwater Plasma Arc Cutting and Secondary Emission Measurements, in Decommissioning of Nuclear Installations, Ed. K. Pflugrad, Elsevier, London, 1990, pp. 208-216.
(58) G. Edler, Master-slave Manipulators for Underwater Applications AKRIBES, Conference on Remote Technology for the Nuclear Industry, London, Dec. 10-11, 1990.
(59) U. Loeschhorn, et al., Decommissioning of the Niederaichbach Nuclear Power Station - Remote Dismantling, Proc. ANS International Topical Meeting, Niagara Falls, USA, Sept. 1986.
(60) J. Griesmeyer, et al., Advanced Handling Technology Project and Implications for Cask Design, Sandia Report SAND-89-0915C, May 1989.
(61) J. Draper, The Human-Machine Interface in Mobile Teleoperators, Workshop on Mobile Teleoperators for Radiological Emergency Response and Recovery, Dallas, June 24-25, 1985.
(62) K. Leinemann, et al., Remote Handling Control with Graphical Man-Machine Interface for NET and JET, IAEA-TECDOC-495, 1989, pp. 215-226.
(63) J. Scott-Hawker, et al., Multiple Robotic Manipulators, Byte, Jan. 1986, pp. 203-219.
(64) J. Draper, et al., Human Factors Activities in Teleoperator Development at the ORNL, Oak Ridge Report CONF-860415-2, Dec. 1986.
(65) J. O'Hara, et al., Control Device Effects on Telerobotic Manipulator Operations, Robotics, vol. 4, 1988, pp. 5-18.
(66) P. Boissiere, et al., An Alternative Control Structure for Telerobotics, Sandia Report SAND-89-0053C, Dec. 1989.
(67) G. Richeux, et al., Préparation du personnel aux interventions, Annales de l'Association Belge de Radioprotection, Vol. 15, No. 1, 1990.
(68) M. Decréton et al., Electromagnetic and Optical Proximity Sensors for In-vessel Handling Unit Control, IAEA Technical Committee Meeting on Robots and remote Maintenance Concepts for Fusion Machines, Karlsruhe, Feb. 22-24, 1988.

(69) T. Raimondi, D. Maisonnier, Latest achievements and Future Needs for Fusion Reactors Remote Maintenance, ORIA-89 Conference on Remote Control in Hostile Environment, Marseille, Dec. 11-12, 1989.
(70) J. Herndon, et al., Advanced Remote Handling Developments for High Radiation Applications, Robots-86 Conference, Chicago, April 20-24, 1986.
(71) P. Drews, et al., Development of Measuring and Control Systems for Underwater Cutting of Radioactive Components, in Decommissioning of Nuclear Installations, Ed. K. Pflugrad, Elsevier, London, 1990, pp. 301-207.
(72) J. Baier, et al., Fernhantierungssysteme fuer die Kerntechnik, Atomwirtschaft, Juli 1990, pp. 351-355.
(73) F. Du Mortier, M. Decréton, Proximity Sensors for Robots in Highly Radioactive Environments, SMIRT-9 Post Conference Seminar on Nuclear Robotics, Lausanne, Aug. 24-25, 1987.
(74) D. Owen, et al., The age of the throw away video camera, Nuclear Engineering International, May 1991, pp. 36-37.
(75) R. Horne, et al., Teleoperation in Radioactive Zones at CERN, ORIA-89 Conference on Remote Control in Hostile Environment, Marseille, Dec. 11-12, 1989.
(76) A. Cregut, An Alternative Scenario for Decommissioning Using Modular Type Vehicle and Telemanipulator, Second International Seminar on Robotics in the Nuclear Industry, Lausanne, Aug. 24-25, 1987.
(77) ICRP, Optimisation and Decision-Making in Radiological Protection, ICRP Publication 55, Pergamon Press, Oxford, 1989.
(78) J. Lochard, et al., Gestion des doses, Préparation et suivi des chantiers, Analyse du retour d'expérience, Annales de l'Association Belge de Radioprotection, Vol. 15, No. 1, 1990.
(79) G. Ferriole, et al., Les perspectives offertes par la robotique, Le programme ISIS, Annales de l'Association Belge de Radioprotection, Vol. 15, No. 1, 1990.
(80) M. Bacot, La préparation des chantiers, Remplacement de générateurs de vapeur, Annales de l'Association Belge de Radioprotection, Vol. 15, No. 1, 1990.
(81) A. Tesini, et al., Experimental Contamination and Decontamination Studies on JET Remote Handling Tools and Materials when Exposed to Tritium, JET Report R(88)18, 1988.
(82) IAEA, Methods for Reducing Occupational Exposures During Decommissioning of Nuclear facilities, STI-DOC-10-278, 1987.
(83) G. Clément, et al., Methodology for Selecting Tooling and Telemanipulated Equipments for a Decommissioning Project, Proceedings of the International Decommissioning Symposium, Pittsburgh, Oct. 4-8, 1987.
(84) V. Massaut, Decommissioning of Research Reactors in Belgium, The BR3 Decommissioning Project, International Training Course on the Decontamination of Research Reactors and Small Nuclear Facilities, Saclay, July 1-2, 1991.

(85) E. Perrott, J. Wakefield, The Development Programme for the Decommissioning of the Windscale Advanced Gas Cooled Reactor, International Conference on Decommissioning of Major Radioactive Facilities, London, Oct. 11-12, 1988.
(86) G. Clar, Nuclear Robotics Technology, Framatome Experience and Objectives, Second International Seminar on Robotics in the Nuclear Industry, Lausanne, Aug. 24-25, 1987.
(87) T. Raimondi, The JET Remote Maintenance Concept, IAEA-TECDOC-495, 1989, pp. 63-74.
(88) Y. Shinohara, et al., Highlights of the Technical Status of Nuclear Robotics in Japan, IAEA-TECDOC-495, 1989, pp. 103-114.
(89) M. Selig, TFTR Maintenance Manipulator, KfK Nachrichten, vol. 18, No. 1, 1986, pp. 18-23.
(90) N. Orlando, A System for Intelligent Teleoperation Research, NASA Report CONF-8310134-2, Dec. 1985.
(91) W. Koehler, et al., Entwicklung hochflexibler Handhabungssyteme, KfK-Nachrichten, vol. 18, No. 1, 1986, pp. 7-12.
(92) B. Houpin, La robotique et les centrales nucléaires, RGN, No. 3, Mai 1987, pp. 296-317.
(93) L. Van Hulle, et al., La robotique appliquée aux interventions sur générateurs de vapeur, Application au manchonnage par soudure laser, Annales de l'Association Belge de Radioprotection, Vol. 15, No. 1, 1990.
(94) R. Vandervort, et al., Remote Cutting and Removal of Chemical Processing Cells's Concrete Pedestal Bases, Proceedings of the International Decommissioning Symposium, Pittsburgh, Oct. 4-8, 1987.
(95) F. Du Mortier, M. Decréton, V. Massaut, C. Hendrickx, Application of Sensory Control in Nuclear Environment, Second International Conference on Intelligent Autonomous Systems, Amsterdam, Dec. 11-14, 1989.
(96) J. Kelsey, et al., Telemanaged Systems, Sandia Report SAND-89-1160C, Dec. 1989.
(97) J. Alfille, et al., Polyjointed Robot with Integrated Laser Beam, in Decommissioning of Nuclear Installations, Ed. K. Pflugrad, Elsevier, London, 1990, pp. 171-180.

MANIPULATORS MASCOT IV USED IN JET AND PROSPECTS OF ENHANCEMENT

T. RAIMONDI, L. GALBIATI
JET Joint Undertaking
ABINGDON
Oxon
OX14 3EA

ABSTRACT The telemanipulator chosen for maintenance of JET tokamak is Mascot IV, a microprocessor controlled unit based on bilateral position servosystem. It evolved from the analogue version Mascot III developed at ENEA in the 60's. The main objective of this type of force feed-back servomanipulator is to give the operator, as nearly as possible, the tactile sensation of actually doing the job. Its success in this depends on a number of characteristics: sensitivity, stiffness, time response and low reflected inertia, to give the operator a prompt and precise feedback so that he can react instinctively. Such characteristics are achieved by using low friction AC motors, high resolution R/D converters, fast sampling time and short delays in the signal transmissions. Velocity and acceleration signals are used to reduce viscous forces and variation of inertia. The digital system has permitted a reduction of the cabling required and the introduction of advanced aids for the operator: teach-and-repeat, useful for precision welding and prepositioning of the Slave arms, toolweight compensation and constraint of the trajectory on given planes or lines. Viewing is assisted by camera tracking.

There is a 16 bit Z8002 microprocessor for each servosystem with 16 I/O, and a servo amplifier. If one servo is out of action the other can still work. Communication between Master and Slave is via a high speed full duplex serial link at 1 Mbaud.

1. Introduction

The telemanipulator is a device which allows an operator to perform a task at a distance, in a hostile environment where human access is impossible or inadvisable. The first developments of such devices were done by Ray Goertz for handling radioactive materials around the end of World War II. The first step was a mechanical through-the-wall master-slave manipulator with direct viewing through shielding windows. Of necessity the work area was limited by the mechanical linkage and the field of vision. The range limit was overcome by the next generation of manipulators operated joint-by-joint. However, precise positioning was difficult and control of forces and

M. C. Becquet (ed.), Teleoperation: Numerical Simulation and Experimental Validation, 139–161.

trajectories were virtually impossible. Electrical master-slave manipulators followed. The slave tongs replicated the movements imparted by the operator to the handles. However the control of forces and compliance with the constraints of the environment were still missing. To reduce the collision forces when the manipulator made contact with the object to be handled, Goertz developed the first force feed-back manipulator in 1948. It was now possible to rotate a handle, brush or grind a surface with controlled forces. The operator could feel what he was doing. Recent developments in the telemanipulator field are the application of computer techniques to ease the work of the operator and supply him with an enhanced feed-back of information. Such techniques are "teach-and-repeat" of the pre-positioning of slave and telecameras in the working environment to save time, constraints of slave movement along preferred trajectories, which is expected to be particularly useful in operations like drilling or cutting, and tool-weight compensation. Research is also going on in the direction of a compact "generalized master" unit, no longer kinematically similar to the slave, which could control various slave units. This will, of course, require kinematic transformations on line and force sensors on the end effectors.

2. Mascot IV

The telemanipulator chosen for maintenance of the JET tokamak is Mascot IV (Fig 1), a micro-processor controlled unit which has evolved from earlier analogue models developed at the Casaccia Laboratory (ENEA) in the early 1960s following in the footsteps of Ray Goertz.[2] In JET the manipulator is deployed in the working areas by large transporters [3].

2.1 FUNCTIONAL REQUIREMENTS

The main objective is to give the operator, as nearly as possible, the tactile sensation of actually doing the job. This is what Jean Vertut referred to as "telepresence" [4]. The manipulator has to be as "transparent" as possible and allow the operator to concentrate his attention on the task rather than on the manipulator.

The dexterity of the servomanipulator depends to a great extent on the following characteristics:

- Sensitivity, defined by the maximum starting load which must be applied to the slave tongs, when the arm is perfectly balanced, to overcome friction and make the servosystem just move. This is essential for controlling the force applied in delicate operations.
- Stiffness, defined as the ratio between the load applied and the consequent displacement between master and slave. This gives the operator the perception of the position of the slave when it comes into contact with the object handled.
- Inertia reflected to the operator, which has to be small so that it does not mask external loads.
- Damping, which has to be high enough to limit the overshoot of the position response to a step variation of load within acceptable values.

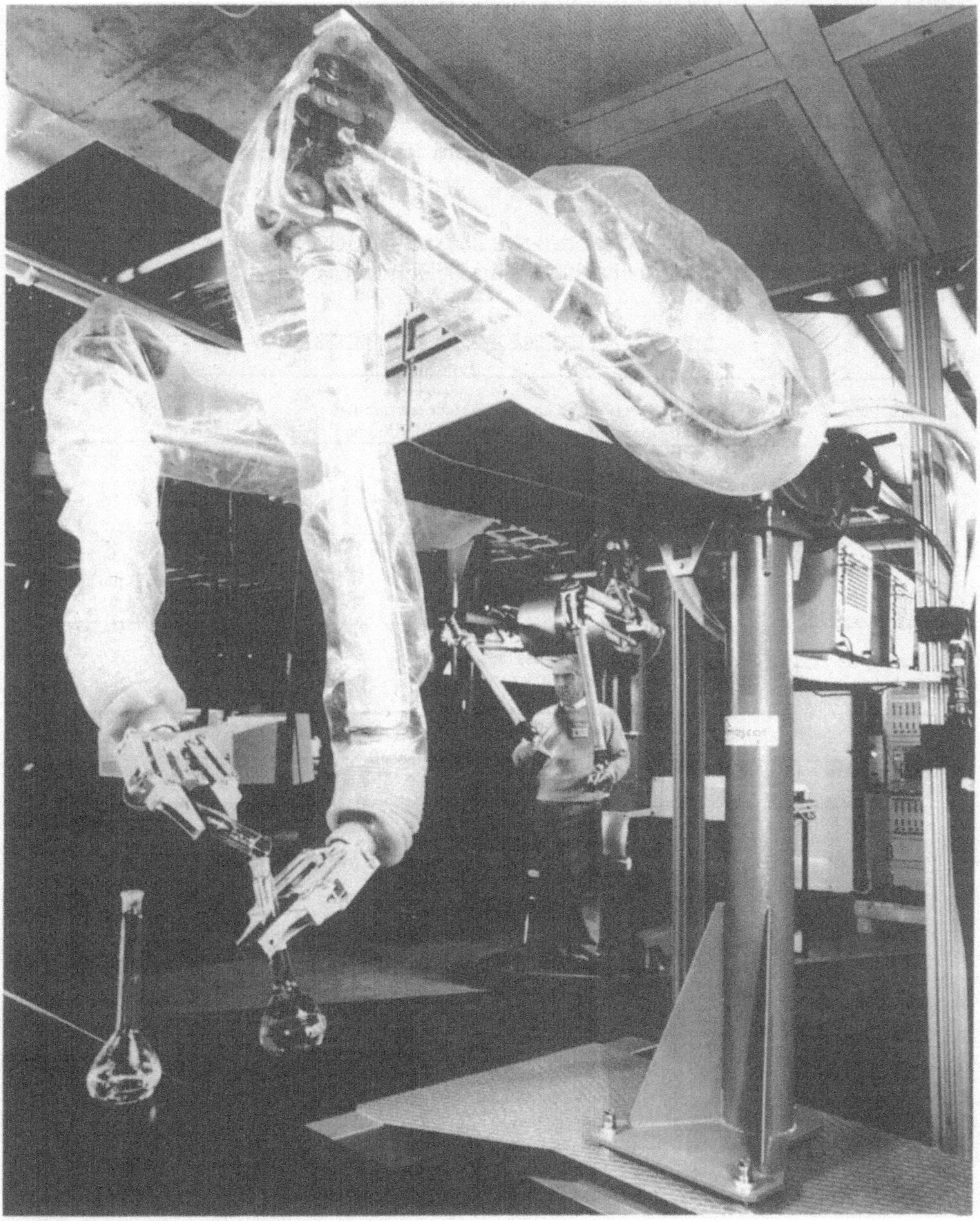

Fig. 1 Mascot IV

- Maximum operating speed. It must be possible to use all the movements freely up to a reasonably high speed without feeling forces which depend on the speed, e.g. viscous damping or the opposite tendency to accelerate, in both no load and full load conditions.
- Time response of the servosystem. It must be fast enough in transmitting variations of forces to the operator so that he can react in an instinctive manner. This is essential to avoid jamming or damaging mating components, cross-threading of bolts, etc.

Table 1 sets out these performance characteristics of Mascot IV, which depend on a combination of mechanics and electronics.

TABLE 1: MASCOT PERFORMANCE CHARCTERISTICS

1 Max. load per arm	20 kg in any direction and any position for ten minutes 12 kg in any direction and any position indefinitely							
2 Gripper squeezing force	24 kg for ten minutes 12 kg indefinitely							
3 Selectable force ratios	Arms: 1:1.5, 1:3 and 1:6 Grippers: 1:1.5, 1:3 and 1:9							
4 Dimensions and weight	Body 405 mm x 860 mm to pass through port One slave arm assembly 110 kg							
	Motions, anotated as in figure 2							
Characteristics ↓	AHEAD X	LATERAL Y	VERTICAL Z	AZIMUTH K	TILT T	TWIST J	GRIP	INDEX
5 Friction -	force ratio 1:1.5 and servomechanisms in operations							
Applied at:-	g	g	g	kg cm	kg cm	kg cm		
Slave	60	80	150	3	2.4	1		
Master	30	30	70	1.35	1	0.5		
6 Stiffness -	displacements with loads applied to slave							
	mm	mm	mm	rad	rad	rad		
3 kg	3	2.5	1					
12 kg	10	8	4					
200 kg cm				0.125	0.25			
25 kg cm				0.04	0.05			
70 kg cm						0.4		
10 kg cm						0.1		
7 Inertia -	reflected to operator with force ratio 1:3							
	kg	kg	kg	kg cm^2	kg cm^2	kg cm^2		
	4	4	9	700	350	100		
8 Damping -	overshoot of final position in response to step in load							
	20%	20%	50%	zero	zero	zero		
9 Speed -	with no load							
	cm s^{-1}	cm s^{-1}	cm s^{-1}	rad s^{-1}	rad s^{-1}	rad s^{-1}	cm s^{-1}	rad s^{-1}
	87	83	83	5.2	10.8	21.6	79	0.052

Computer-aided functions, which were pioneered by Jean Vertut, are also important as aids to the operator:

- Teach-and-repeat. This is useful for repetitive operations or where viewing is impaired, eg repair welding. It is also used for the transporter which positions the manipulator in the work area.
- Tool-weight compensation. This is particularly important when the slave has to work in a different attitude from that of the master. It also alleviates operator fatigue while allowing good sensitivity.
- Constraint of the trajectory on given planes or lines. This is useful, particularly where viewing is not good, to keep a tool on the line or plane of action. Operations such as sawing, screwing and aligning connectors can be greatly facilitated by this technique.

The viewing system must be integrated with the telemanipulator. The operator will be disoriented unless the telecamera is in approximately the same attitude with respect to the slave tongs as the operator's eyes are with respect to the master handles. Depth perception can be provided either by stereoscopic cameras or by an auxiliary camera to supply side views. High resolution is essential.

2.2 MECHANICAL CONFIGURATION

The manipulator has two arms with six degrees of freedom, seven if we count the squeezing motion of the tongs. The arms are independent so that one can be removed if single arm operation is desired. The geometry of an arm is shown in Fig 2. All the motors are located on the shoulder disc except the one which actuates the roll of the arm (X motion). This serves as a partial counterweight for the arm. The advantage of situating the motors on the fixed shoulder rather than on the joints is lower inertia. In addition there is no torque "cross coupling", i.e. the reaction to the torque exerted by one degree of freedom (d.o.f) is taken up by the shoulder and not by the other d.o.f (s). The disadvantage is a bigger shoulder and lower stiffness on the wrist (J,T,K) motions which are transmitted via steel wires. The arm and forearm are not demountable. As the range of X and Y motions is rather limited it becomes possible to transmit the Y motion with a simple articulated parallelogram. For telemanipulators, which are positioned in the working area by other devices, the range of the servosystem is not of such importance as for industrial robots. Dynamic cross-coupling between X,Y and Z is also limited by the limited range of motions (Fig 2).

The X,Y,Z motions are counterbalanced in all attitudes by mechanical counterweights (Fig 3).

In Fig 4 the etched areas show the working volume covered by the servosystems of the slave arm. An indexing motion is provided on each arm so that the working volume of the servosystems can be displaced over a range of 180°. In this way the overall angular range of shoulder motion is 270°. The shoulders are mounted on the terminal part of the transporter (boom

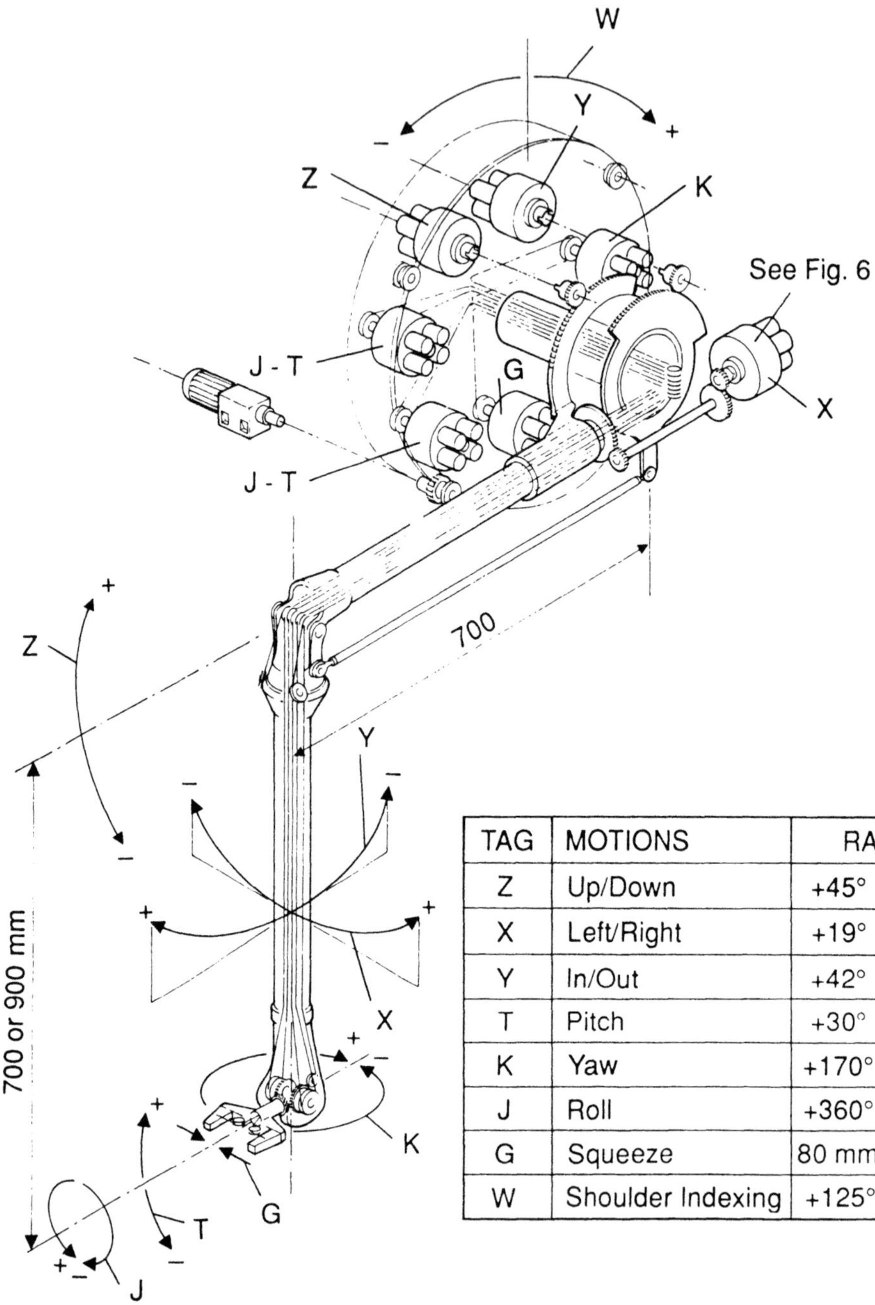

TAG	MOTIONS	RANGE	
Z	Up/Down	+45°	–45°
X	Left/Right	+19°	–45°
Y	In/Out	+42°	
T	Pitch	+30°	–120°
K	Yaw	+170°	–170°
J	Roll	+360°	–360°
G	Squeeze	80 mm	
W	Shoulder Indexing	+125°	–55°

JG91 516/19

Fig. 2 Geometry of Slave Arm

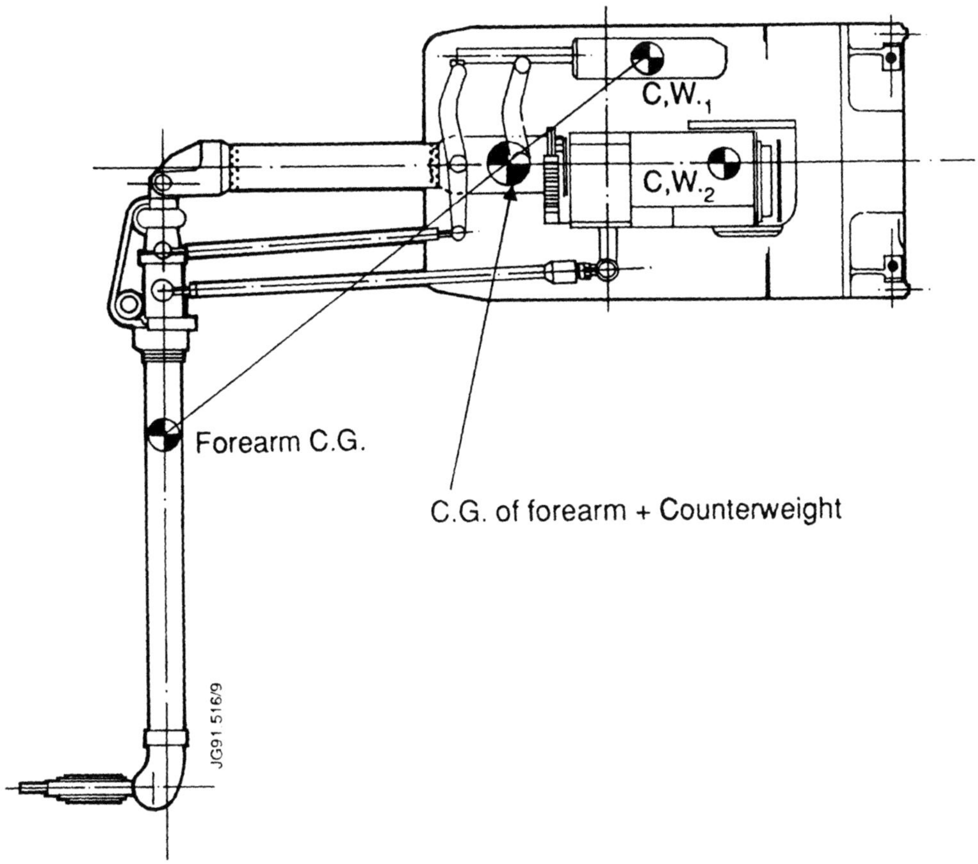

Fig. 3 Counterweights of Arm and Forearm

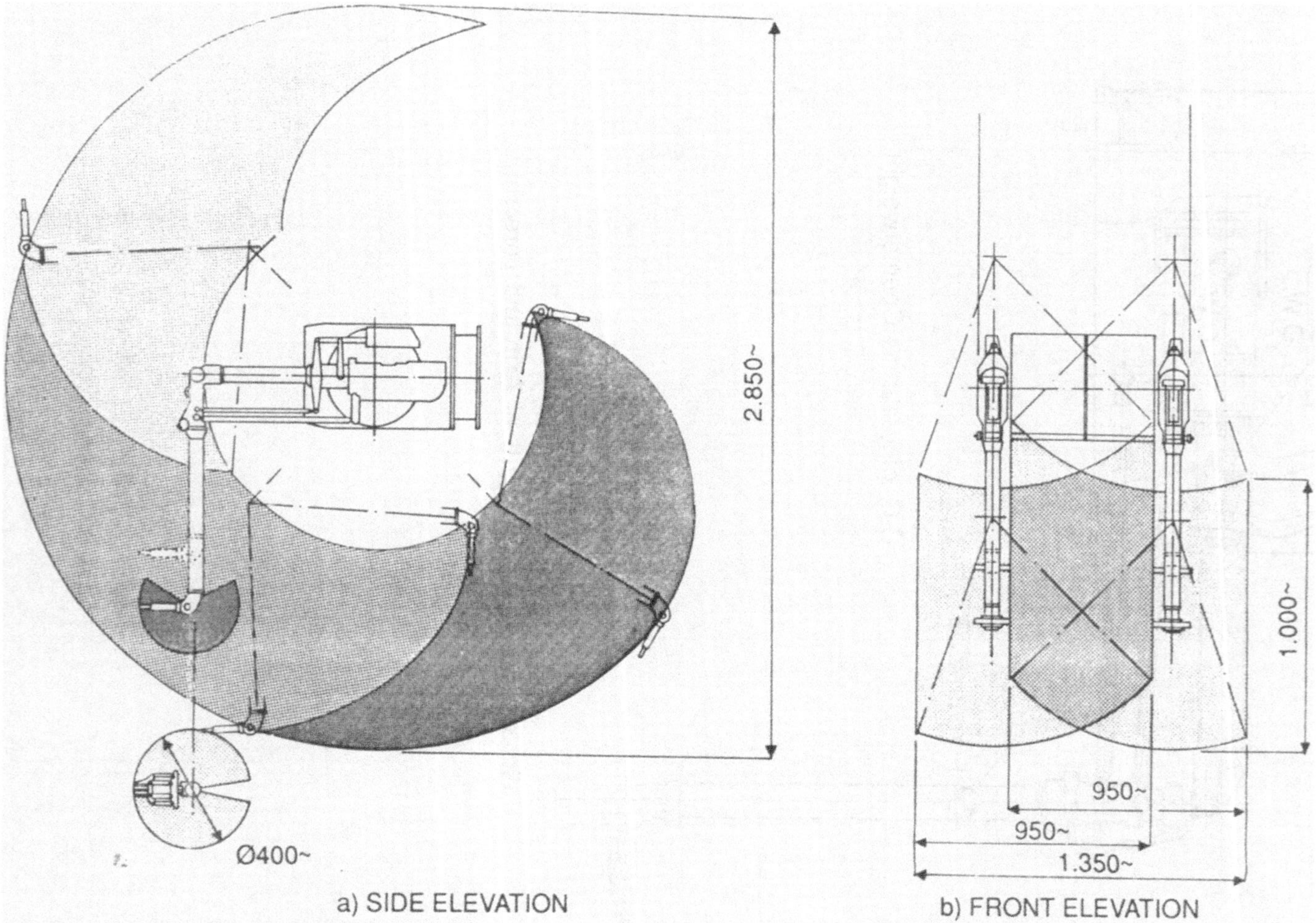

Fig. 4 Working Volume of Mascot IV

extension) which is provided with pan/tilt/roll motion to further extend the working volume of the manipulator and to reach the tool box (Fig 5).

In Fig 5 the layout of the television system is also shown. One telecamera is installed on the front of the manipulator. It has 4 d.o.f and virtually acts as the eyes of the operator. A second camera mounted on the manipulator support provides side views which the operator refers to at intervals to get depth perception. It is instinctive for the operator to move his hands parallel to the direction he wishes the motion to assume on the screen. So there must be consistency between the visual feed-back and his action. Ideally the operator's line of vision to the monitor should be in the same direction with respect to the master arms as the optical axis of the camera with respect to the slave arms. In practice deviations of up to 45° from this ideal are tolerable. The monitors are suspended from a rail so that they can be moved around the operator and their height can be adjusted. For many tasks the attitude of the slave will be different from that of the master. It may have to work upside down or on its side. In these cases the gravity reflected to the operator is not in the vertical direction. The operator soon gets used to this but it will be very useful in this case to cancel the weight of the tools with the algorithm mentioned above.

2.3 STRUCTURAL DESCRIPTION

Six actuators are mounted on a cast aluminium alloy disc which houses concentrically the bearings of the Z and of the Y axes. The Y motion is transmitted to the forearm by a connecting rod in an articulated parallelogram arrangement. The disc is supported by four rollers and can be indexed by another actuator (not force reflecting) via a spur gear. To minimize dimension the rollers are housed in a thin aluminium box (Fig 6). In the areas marked s the material is reduced to a 2mm membrane which can only resist forces in its plane, while stability of the box is provided by the resistance to compression of the disc itself along the diameter passing through the rollers (Fig 6).

The actuator for the X motion is accommodated at the back of the arm and acts as a partial counterweight of the Z motion, supplemented by tungsten plates. The X and Y motions are balanced by tungsten counterweights mounted on an articulated parallelogram arrangement in the same plane as the arm and forearm. As shown in Fig 3 the resultant gravity force due to the counterweight CW1 and the forearm is applied to the arm in a fixed position. This force, summed to the weight of the arm, is balanced by CW2. Mechanical counterweighting was chosen, rather than an electrical method, to avoid overheating the motors. The greater inertia and overall weight were considered less of a drawback and inertia can be reduced with control algorithms.

Arm, forearm, and connecting rods are in high-strength aluminium alloy, while wrist and elbow are in aluminium-magnesium castings. The wrist motions are achieved with a differential arrangement moved by steel wires and pulleys. All bearings are in stainless steel.

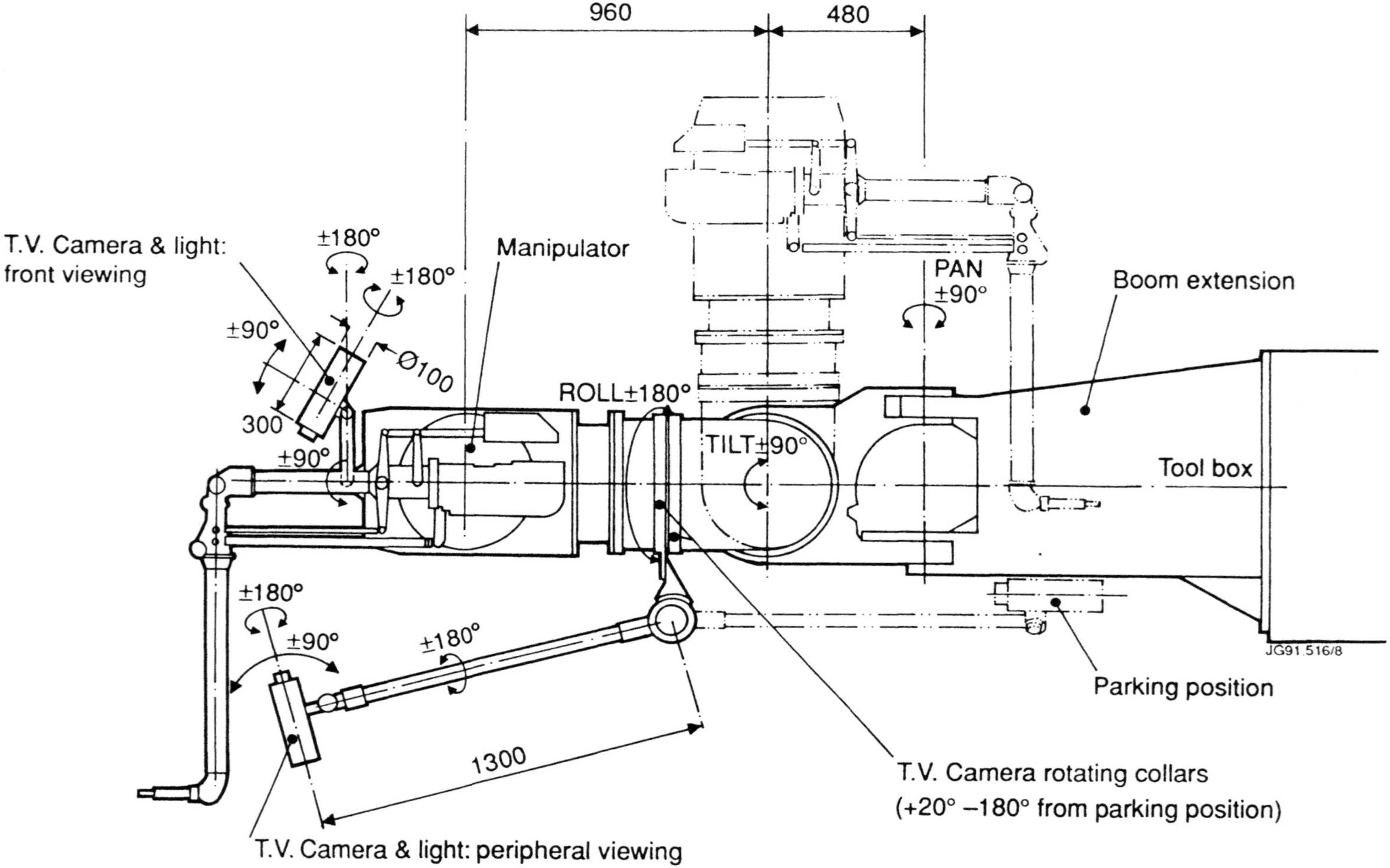

Fig. 5 Manipulators and Cameras on Boom Extension

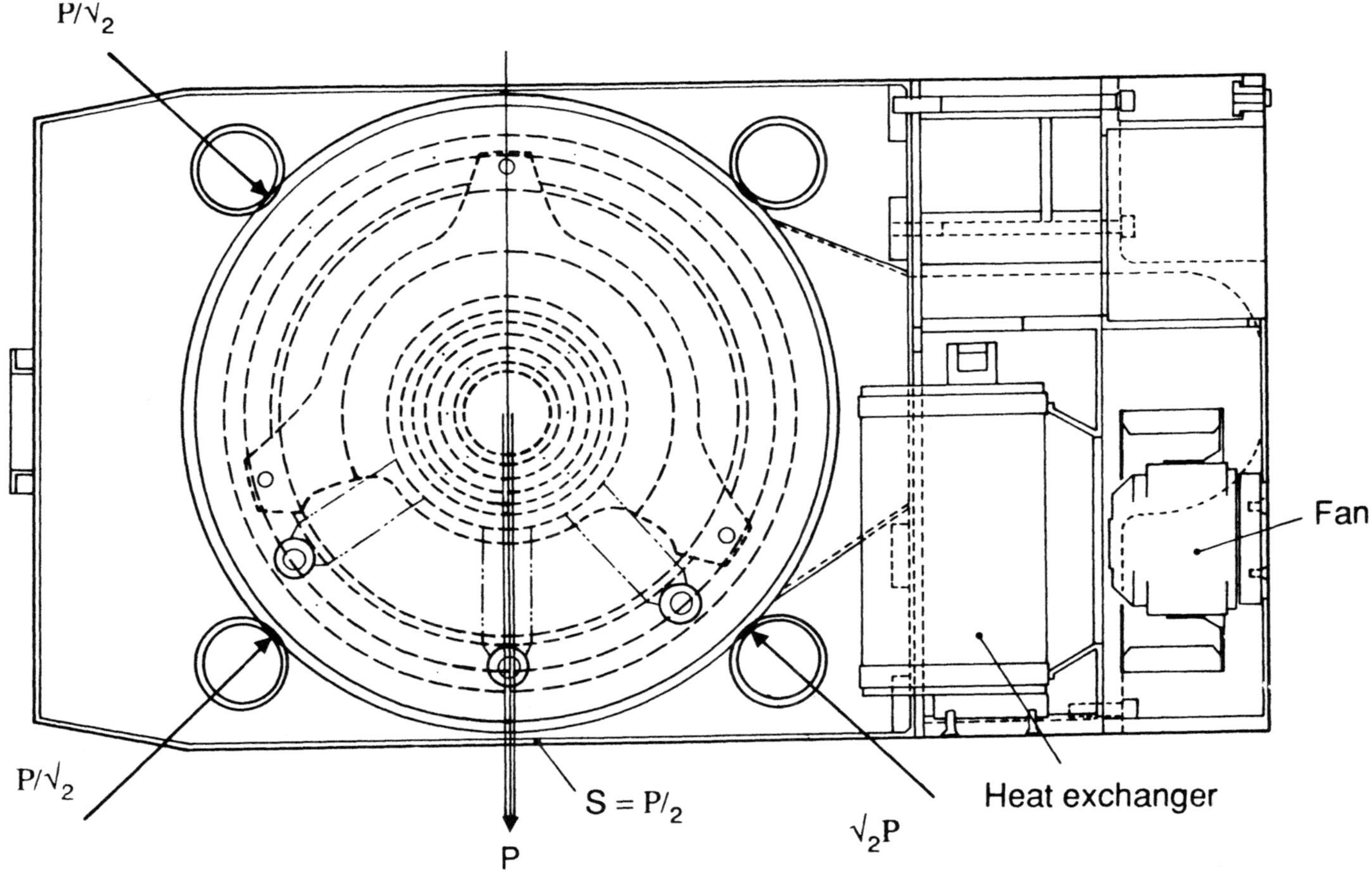

Fig. 6 Forces in Containment Box

Closed air circulation is essential to avoid on one side, ingress of contaminated air into the manipulator, and on the other outflow of air from the manipulator to avoid polluting the environment. This means that a water-to-air closed loop heat exchanger is housed at the back of the manipulator, together with a fan which blows onto the motors.

2.4 ACTUATORS USED FOR MASCOT

Two types of actuators are used, one for the master and one for the slave. Seven are required for each arm. The slave actuator (Fig 7) consists of a gear box with three-stage precision spur gears, offering the required high efficiency, with a 40:1 gear ratio for the slave and 25:1 for the master. Special attention has been given to gear material. Pinions in Martensitic stainless steel, heat hardened to 65 Rc, engage wheels in similar material or in high strength anodized aluminium alloy. This choice aims for long life, low friction and inertia and washability.

AC two-phase induction servo motors (four on the slave, two on the master actuator) are used, working at 60Hz. Friction torque is low, so that high sensitivity can be achieved, and there is no cogging torque, which favours smoothness of operation. The motors chosen have a high power ratio (T2/J), and offer a larger surface for heat exchange than an equivalent single motor. They are designed to operate with a fixed voltage applied to the reference phase and variable voltage to the control phase. They have considerably resistive rotors to provide almost linear torque/speed characteristics. The main shortcoming is the problem of heat generation and overheating of the windings. However, insulation resistant to higher temperatures is now available. Another shortcoming is that if only one of the two phases is controlled the motors give a "buzzing" torque at 120Hz which is a nuisance for the operator. This is eliminated by controlling the two phases at the same time but the torque becomes a quadratic function of the input signal and this necessitates a square root compensating network. In the present unit a compromise has been reached by having a linear control on both fields plus a constant basis on one of the two. The position transducers used are resolvers geared down to give the absolute position of the actuator. One revolution covers the full range of motion. The resolver and gear head is connected to the first wheel engaged by the pinions of the motors. Backlash in this coupling is eliminated by preloading the resolvers with a clock spring. Backlash between wheel and pinion is virtually neglible thanks to the compliance of the plastic material used for the wheel and accurate machining. The relative weakness of the material is compensated by sharing the load onto four pinions. The actuator has a fail-safe brake consisting of a latch lifted by a solenoid and released to brake the first wheel when power is cut or alarm signals are given by the control system. The braking torque is limited by a clutch so that in case of failure of the servosystems it is still possible to retrieve the manipulator.

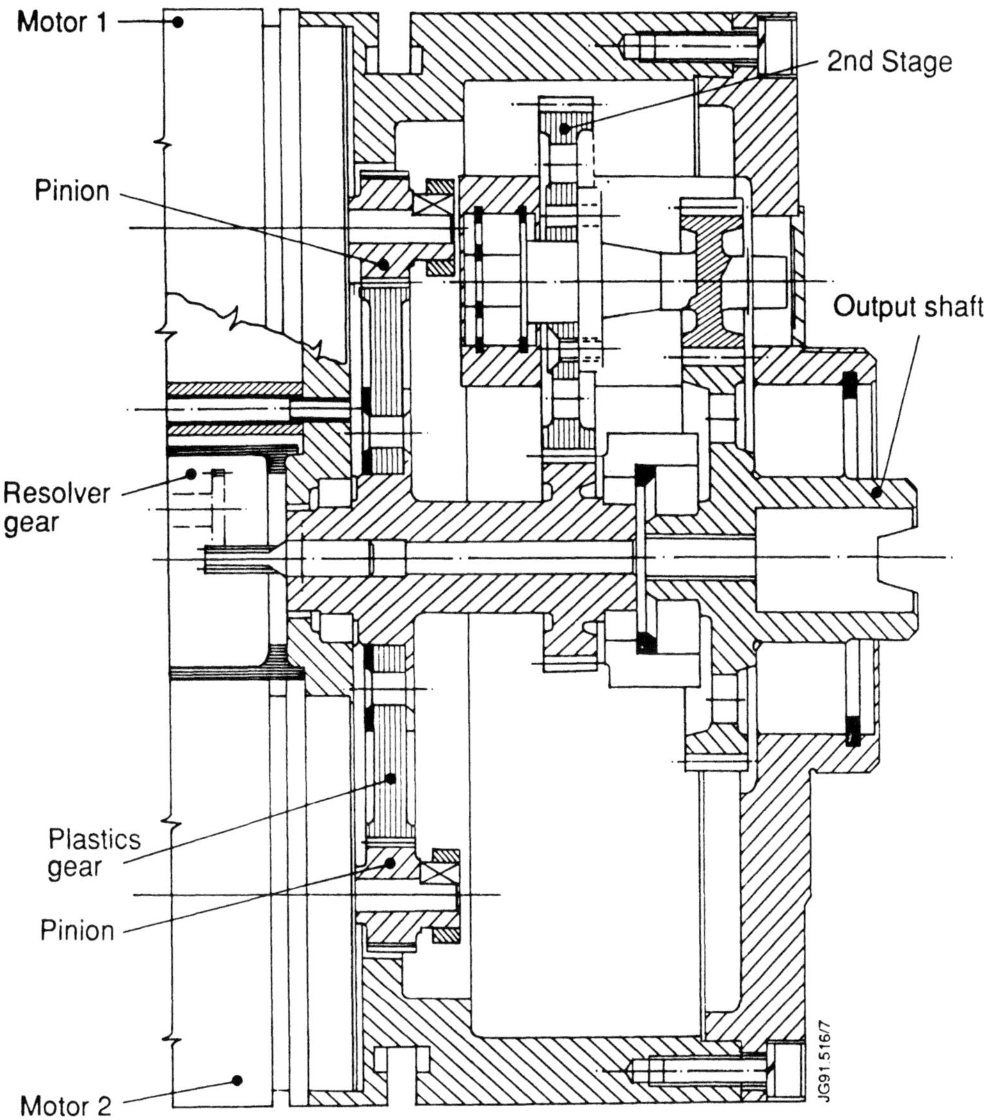

Fig. 7 Mascot Actuator

2.5 ALTERNATIVE ACTUATORS

Present day microprocessor technology permits heat losses to be reduced by a combined frequency and current or voltage control. In this way the slip, which is a major factor of thermal loss, can be limited and stator current (or voltage) can be computed on line to achieve the desired linearity between torque and input signal. The first manipulator controlled by this technique has recently been developed under the auspices of KFK, Karlsruhe. 3-phase motors have been adopted to avoid the third harmonics of the magneto-motive force distribution. Frequency-voltage control has been recently tested on the Mascot, indicating a possible reduction of the power losses by 20% [5].

Another actuator which might be considered is the brushless synchronous type. The principle, shown in Fig 8 is that the magneto-motive force generated by the stator rotates synchronously with a permanent magnet rotor and an angle of 90° to the flux to maximize the torque. The MMF is synchronized to the rotor by using a resolver mounted on the shaft of the motor, which can also be used as a position and velocity transducer [6]. This type of motor, with rotors in the materials with high magnetic energy now available (rare earths), achieves the maximum power/weight ratio. There are however, two drawbacks for its application to force feed-back teleoperators: a drag torque caused by magnetic hysteresis in the iron of the stator, and a "cogging" torque due to the ripple of the magnetic force applied to the rotor by the teeth of the stator. The drag torque could be compensated by positive velocity feed-back of constant amplitude (cf. para 3.2). "Cogging" torque can be compensated by skewing the teeth of the stator, by using odd numbers of slots and larger gaps, and where there is more than one motor, mounting them in pairs with one stator turned by half a tooth pitch so that the torque ripple of one is compensated by that of the other. Cogging and harmonics ripple torque can be furthermore reduced by injecting a current value depending on the angular position of the rotor, e.g. by means of "look-up" tables. Further investigations of the possibilities of brushless motors for teleoperators are in progress within the E.C. Teleman programme.

DC motors have been used for teleoperators. An obvious advantage is that they require only one amplifier. The main drawback is friction and wear of the brushes. Starting friction cannot be compensated with positive velocity feed-back since direction of motion is not detectable without force transducers (cf. para 3.4).

3. Control System

3.1 THE FORCE FEED-BACK CONCEPT

The block diagram in Fig 9 shows the control concept. It is a bilateral position servosystem. The position demand applied to the slave is the actual position of the master and vice-versa. For both master and slave, the position demand is compared to the actual position and the error is transformed into a proportional stall torque command applied to the motor. For the slave, this torque is summed to the load torque T2 and to the friction torque CA2. The result, multiplied by the transfer function of the actuator, containing the

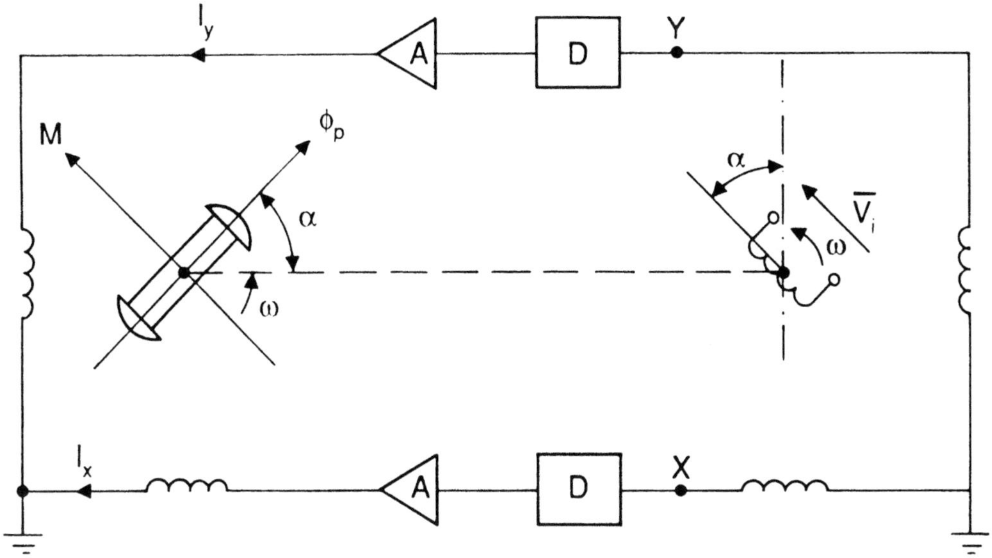

Fig. 8 Brushless Motor with Resolves

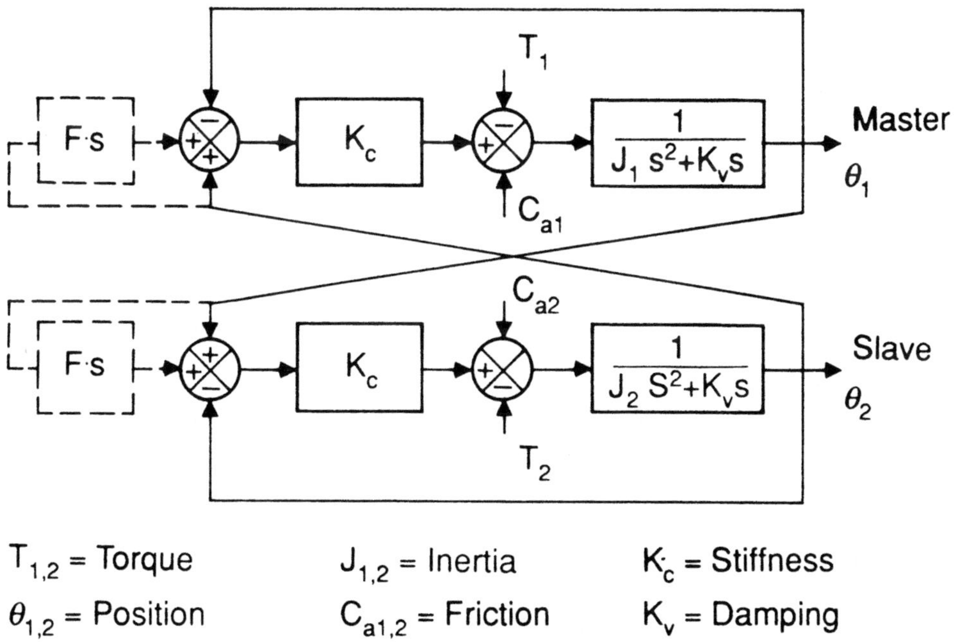

Fig. 9 Force Feedbacks Concept

inertia J2 and the damping factor Kv, gives the angular position θ_2 of the actuator. The damping factor Kv is the natural damping of the motors plus the signal generated by a tachometer. This is equivalent to the mechanical scheme of Fig 10. For the master, the same happens but the torque T2 is replaced by torque T1 applied by the operator. Fig 11 is a scheme of the whole bilateral servo composed from Figs 10a and 10b. At steady state the torque applied by the operator to the master is equal to that applied by the slave to the workpiece, or vice-versa as the system is symmetrical. This torque is proportional to a small displacement $Kc(\theta_2-\theta_1)$. The quantity θ_2 is the only essential piece of information which must be transmitted from slave to master. The higher Kc, the higher the stiffness of the transmission.

From Fig 11 it is clear that if the operator imposes a constant speed $\dot{\theta}_1$ on the master side, the slave will only experience a torque T1 - Kv $\dot{\theta}_1$, and the torque T2 will only be T1 - 2Kv θ_2. If vice-versa, the damping was produced by the quantities $Kv(\dot{\theta}_2-\dot{\theta}_1)$, then there would be no viscous effect after the initial transient, that is when $\dot{\theta}_2 = \dot{\theta}_1$. This is in fact what is achieved in the analogue version of Mascot where it is the difference of the two speed signals that is applied to the servos. In Mascot IV the same effect is obtained by adding velocity feed forward signals. The equivalent mechanical scheme is the one in Fig 12. By analogy with electrical circuits, we can consider this as a mechanical quadrupole. Defining the impedance as the ratio of torque and speed, we can speak of operator with high impedance when he imposes movements not affected by the torque applied, or low impedance when he imposes forces not affected by the velocity.

If the system is linear, from the simple analysis above one can deduce that there is asymptotic stability, since the system is equivalent to a passive quadrupole, at least when the same parameters Kv, Kc are used for master and slave. The same result can be obtained even if $Kc1 \neq Kc2$ and $Kv1 \neq Kv2$, by evaluating the solutions of the characteristic equation of the closed loop system, using for instance the Routh criterion. However, there are other reasons neglected before which, in practice can make the system unstable when the gain Kc is increased beyond a certain limit. These are: electrical poles, elasticity of the joints, and non-linearities such as backlash, delays saturation. In practical terms, if Kv is reduced below certain limits, the damping is reduced and low-frequency oscillations tend to start. If Kv is too high, the system will tend to oscillate at high frequency due to the increased bandwidth of the velocity loop.

3.2 CONTROL ALGORITHM

The control algorithm is shown in Fig 13. The position signal of the master is sent to the slave with a sampling time of 2.5ms. It is compared to the position assumed by the slave at the same instant. The error is fed into a velocity loop as a speed demand and compared to the velocity signal supplied by a 400Hz tacho generator sampled at its peaks, that is every 1.25ms. The velocity error is then compared to the acceleration derived as increment of the velocity peaks and the difference is amplified, modulated at 60Hz and applied to the control phase of the motors. The absolute value of the same signal, summed to a preset basis, and phase shifted by 90°, is applied to the fixed phase. Each

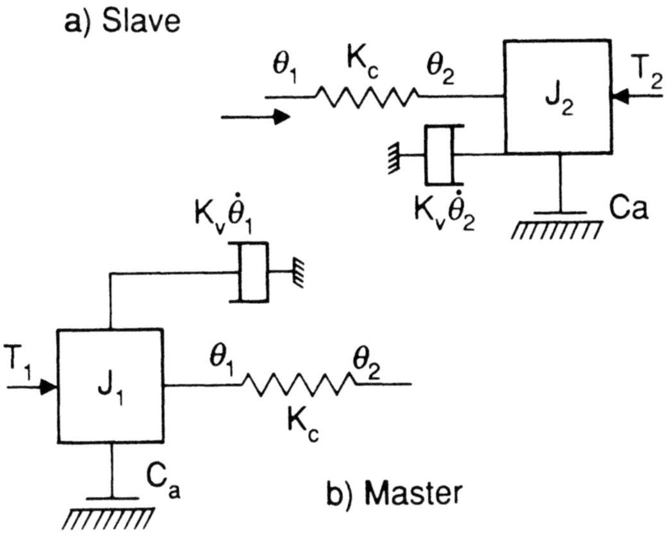

Fig. 10 Mechanical Equivalent

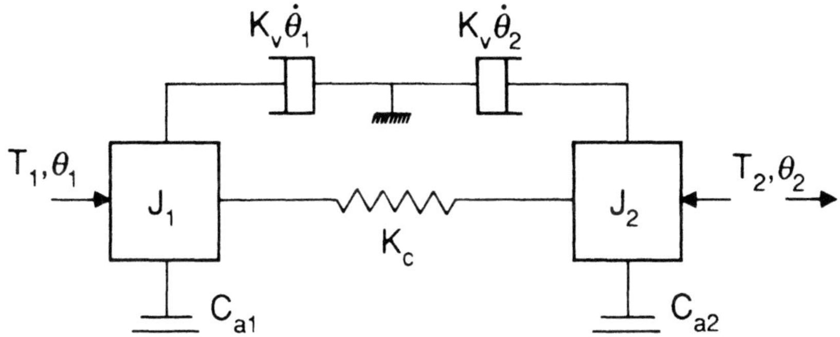

Fig. 11 Bilateral Servo Without Velocity Feed-forward

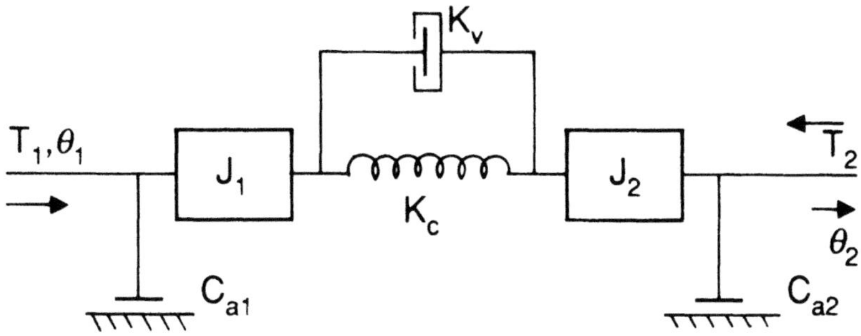

Fig. 12 Bilateral Servo With Velocity Feed-forward

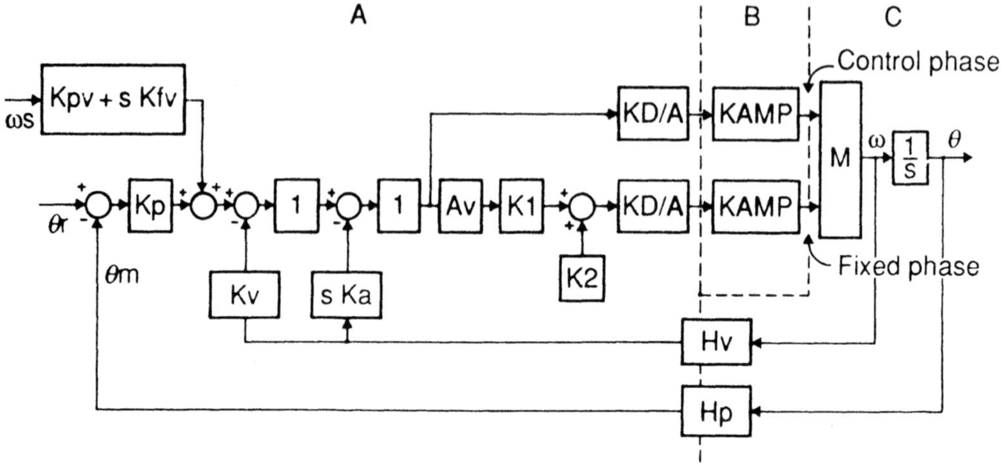

θr	=	position input signal
θm	=	position feedback signal
Kp	=	position loop gain
Kpv	=	feed forward gain
s	=	Laplace operator
Kv	=	velocity loop gain
Ka	=	acceleration loop gain
Av	=	this function gives on its output the absolute value of the input signal
K1	=	this parameter determines the relation between fixed and control phases of the AC servomotor
K2	=	basis on the fixed voltage
KD/A	=	gain of the D/A converters
KAMP	=	gain of the power stages
M	=	AC servomotor
ω	=	angular speed of the motor
θ	=	angular position of the motor
Hp	=	position feedback gain
Hv	=	velocity feedback gain
Ha	=	acceleration feedback gain
Kfv	=	acceleration feedforward

Fig. 13 Control Algorithm (Masters)

servo has two PWM type power amplifiers pulsed at 20KHz, a rather novel application of the switching mode to an AC system aimed at reducing heat dissipation. The velocity signals provide a velocity feed forward which compensates the viscous drag. In this type of application the inertia of the load varies from zero, when the manipulator is moved with no load, to virtually infinite when the manipulator tongs are pressed against a firm wall. The inertia of the operator is also variable. The inner acceleration loop has the effect that the acceleration achieved tends to be independent from the inertia of the load, and therefore an optimisation of the response is possible for all the load conditions. The frequency bandwidth in no load conditions is 20Hz for both master and slave servos.

3.3 CONTROL ARCHITECTURE

The control equipment consists of the Master cubicle and the Slave cubicle. Each contains two independent modules, one for each arm. Each module contains one 16-bit microprocessor for each servosystem, connected to a local bus. If one servo is out of action the others can still work. The modules are connected via a common bus through a full duplex serial link at 9600 bps. The communication between each Master module and the corresponding Slave module is accomplished by means of a high speed full duplex serial link at 1Mps. This link has a range of up to 2km. A portable PC is the interface for setting parameters and monitoring the system, which is provided with on- and off-line diagnostics.

3.4 ENHANCEMENTS

Reduction of inertia reflected to the operator is possible with acceleration feed forward.

Some granularity was present due to the digital derivation of the position. This initially improved transmitting and derivating the tacho generator signals from the slave to the master and vice-versa, but sampling rate must be increased to obtain a worthwhile reduction of the inertia reflected.

A further improvement of sensitivity could be obtained by the addition of torque or force transducers on the output shaft of the actuators. The difference of the two torque values drives the two actuators in the same direction so as to balance the difference itself. In this way, not only the positions of master and slave follow each other, as described, but the torques applied by the actuators must be the same. This helps in reducing friction and inertia in the movements of the arms, due largely to the servo drive units, and improves the accuracy of the transmission of forces, under both static and dynamic conditions, that is the "feel" of the manipulator (Fig 14). However instabilities may occur when the slave is against a firm constraint. They can be compensated by P, D, algorithms on the force signals and by making use of the compliance of the arm.

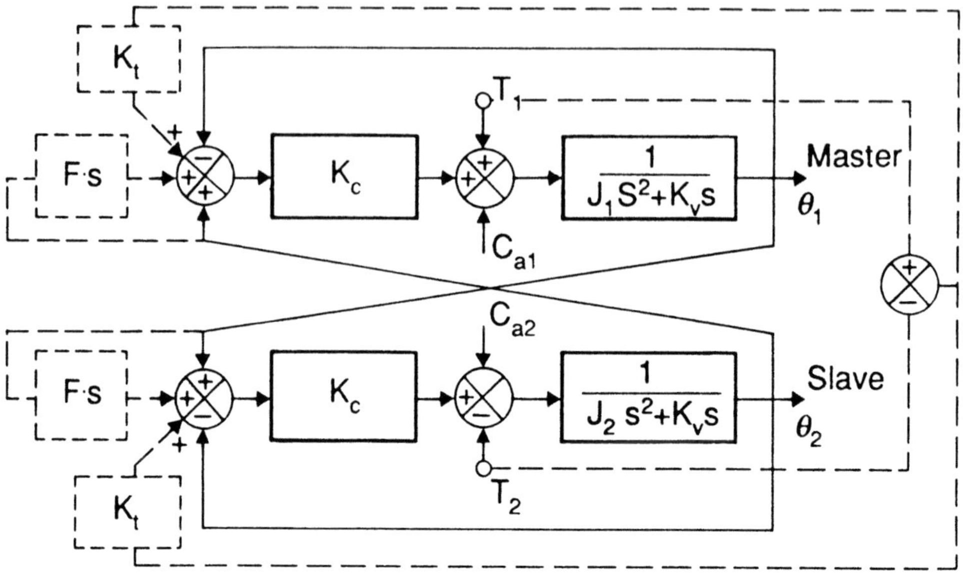

$T_{1,2}$ = Torque $J_{1,2}$ = Inertia K_c = Stiffness

$\theta_{1,2}$ = Position $C_{a1,2}$ = Friction K_v = Damping

Fig. 14   Bilateral Servo with Torque Transducers

4. Advanced Functions: Computer Aided Telemanipulation (CAT)

The principal aids to the operator, as mentioned in para 2.1, are the following:

- Teach and repeat.

In order to reduce the memory requirements a parabolic approximation of the trajectory is used in the following way.

During teaching, eight position samples per d.o.f are sent every 20ms from the Master resolvers to the auxiliary CPU. The first and last are stored in the RAM. Using the first eight samples and imposing zero derivative velocity, a parabolic function is calculated as a best fit by the CPU and the difference between this parabola and the 8th sample is detected. If this is less than a present threshold (3 bits, equivalent to 0.1mm) then the process is repeated but using 11 samples instead of eight in the next period, at the end of which a comparison between the parabola and the actual 11th sample is made. If the error is less than a present threshold, then the next phase will include more samples, if it is larger than a second threshold it will include only eight samples. These will be used to calculate a new parabola fulfilling the condition of derivative continuity with the old one. Tests have been done teaching and repeating at various speeds, with the master holding a ball point pen to draw irregular shaped lines and repeat them. Repeatability was estimated at < 0.2mm within the range of motion where backlash is recovered by gravity. This augurs well for the feasibility of remote repair welding, with the addition of arc voltage control.

- Constraints on given planes

The method is the following. Starting from the angular positions of the slave d.o.f's the first step is to evaluate the cartesian coordinates of the slave wrist and then of its projection on the constraining plane. These are then transformed into the corresponding angular positions which are fed into the Master. The slave follows in the normal way. If the slave experiences a force that tends to displace its wrist out of the desired plane, the master wrist will remain on the plane and tend to bring the slave back to it. The operator will retain force feed-back and freedom of motion in all directions except the one perpendicular to the constraining plane. If he deviates from this, he will feel a force opposing him, the stiffer the servo the more sharply the force increases.

- Tool weight compensation

The torques at the various joints, proportional to the angular displacement, in a particular configuration of the slave give, at steady state, information on the weight of the tool. By trigonometric calculations it is possible to compute on line the balancing torques needed at the various joints to compensate the weight in any configuration.

- Observation

"Telepresence" is the goal and it cannot be achieved unless the viewing system is able to give the operator of the remote manipulator global 3D information on the remote handling area. This goal is at the moment far from realization though development work is being done. Some lines that need to be pursued are:

- Real time simulation and graphical overlay

  Often only a partial view is available and important details are either hidden or poorly illuminated. The objective is to overlay on the TV picture a graphical model which completes the picture, and can give 3D clues in the absence of a stereo camera.

- On line image processing for dark areas or to reduce "noise"

- Autofocus and autoiris with manual option

- Aiming of the various TV cameras

To make this easier the prepositioning of the cameras should done through teach and repeat. Where this is not possible, the operator should be able to aim at a location with any conveniently placed cameras by moving a cursor on a graphical model of the machine or of the part to be worked on. He should also be able to move the support of the cameras in a similarly "friendly" way. To avoid disorientation of the operator computer aids need to be developed to orient the cameras, in particular the roll motion, consistently with the displacements between master and slave units. Cameras should also be able to track automatically a selected part of the manipulator, object handled, or transporter while they are moving. KFK is developing a viewing control system for JET incorporating these features.

Image processing will be used to compensate oscillations. During work oscillations may be imparted by forces on the slave or end effector. Low frequency oscillations do not appear to adversely affect the operator. Higher frequencies (above 1Hz) would be of some concern, but mounting the TV cameras on a soft support would attenuate them should the low frequency oscillations prove to be a nuisance in the long run. Present day image processing techniques could be applied to stabilize the image.

Acknowledgments

The authors wish to acknowledge the vital contributions to this project of the JET Remote Handling Groups, in particular R. Cusack for lay out and mechanical aspects, of Prima Progetti and Elsag SpA for design and manufacturing, and of S. Colombi and G. Costi for improvements of the control system.

References

[1] C. Mancini, F. Roncaglia, Il servomanipulatore Electronico Mascot 1 del CNEN, Alta frequenza No. 6-1963.

[2] L. Galbiati, C. Mancini, T. Raimondi, F. Roncaglia, A Compact and Flexible Servosystem for Master-slave Electric Manipulators, 12th Conference on Remote System Technology A.N.S. 1964.

[3] T. Raimondi, The JET Experience with Remote Handling Equipment and Future Prospects, S.O.F.T. - Utrecht 1988.

[4] J. Vertut and P. Coiffet, Teleoperation and Robotics, Robot Technology Vol. 3A, p.45.

[5] S. Colombi et al, Improvements of Actuators for Servomanipulators, O.R.I.A. 91 - Marseille: Proceedings to be published.

[6] C. Mancini, T. Raimondi, Patent No. 891506 (1/10/71).

POSITION SENSING FOR ADVANCED TELEOPERATION IN NUCLEAR ENVIRONMENT

Marc Decréton
CEN/SCK Nuclear Research Centre
Boeretang, 200
B-2400 Mol
Belgium

ABSTRACT. Advanced teleoperation requires an autonomous perception capability of the remote operated equipment. This perception complements the vision information provided to the operator by cameras and allow for computer controlled task execution. An important part of this perception is obtained through position sensors. However, for nuclear applications, severe environmental restrictions, in particular w.r.t. radiation, limits the choice of such components. This paper discuss the specific requirements for these sensors, the possible solutions and their availability. Particular attention is put on the internal position sensors of the teleoperated machines, and the non-contact distance measurements to be performed at the end effector level.

Part of this work has been performed under contract No. FI2T-CT90-0013 (VISYS) of the European Teleman Programme.

1. Introduction

The use of advanced features in nuclear tele-manipulated tasks in order to get a high level of autonomy presents a particular challenge to the present robotic technology. Nuclear applications have generally particular requirements, different from typical industrial fabrication processes:

- the manipulator must work in a highly **unstructured environment** not specifically constructed for robotised machines,
- no full autonomy is feasible with the present state-of-the-art technology, so that **mixed control mode** must be used, with both purely manual operation, and specific autonomous tasks,
- a high level of reliability is needed, as usually man-access to the machine is difficult. A certain level of "self-rescuing" or at least of "**remote rescuing**" capability in case of failure is required.

M. C. Becquet (ed.), Teleoperation: Numerical Simulation and Experimental Validation, 163–194.

- **environmental** hazards are high and offer potential **degradation** danger for the performance of many standard components. This concerns not only nuclear radiations, but also in specific applications, high temperature or pressure, underwater work, presence of toxic gases or liquids.

Nuclear applications of autonomous manipulators lead therefore to a particular paradoxical challenge:
- to require a **high level of sensing and intelligence** to cope with a partially unknown and unstructured environment,
- and at the same time, to require such a **high degree of reliability** and safety under the hazardous environment present, that many sophisticated features of robots must be excluded.

This is particularly true for the aspects of sensing. Sensing is essential to acquire a good perception of the environment and therefore to act in an autonomous way. But many sophisticated sensors are too sensitive to be used in a nuclear environment, and see their performance degraded, or even destroyed, after a given operation period.

This paper will discuss the basic requirements put on sensors to be used on semi-autonomous manipulators in nuclear environment. The discussion will be concentrated on position sensors, but the general analysis is also valid for the other measurements, such as force, tactile sense, etc. It will also limit its scope to non-vision sensors. Moreover, the main application in mind will be cell type tasks using a semi-autonomous gantry crane system. Such installations are typical of many applications in decommissioning or reprocessing tasks. Typical problems of mobile platforms, particularly the long range navigation sensors, will not be considered here in full details.

Chapter 2 will review the general requirements on position sensors for semi-autonomous teleoperation. Specific environmental requirements are dealt with in Chapter 3. Chapters 4 and 5 will review the existing techniques for internal sensors and non-contacting position sensing respectively. The last chapter will present a short list of sensor types suitable to be used in nuclear applications, together with the needed assessment work still required.

2. General requirements on position sensing

2.1 Needed data

In semi-autonomous teleoperation, an intelligent manipulator is used to perform tasks remotely, in such a way that operator control is greatly facilitated by specific features of the machine used.

Primarily, this can be obtained by two capabilities (Fig. 1):

- the **position and interaction** of the manipulator, both with respect to its base and with respect to its surrounding, is known with sufficient accuracy by appropriate measurements. These data are essentially used for two purposes: to navigate up to the goal position in a safe and efficient way (navigation means sensing over large distances with narrow field of view, safe means to avoid collision, i.e. to sense obstacle close-by with a large coverage) and to perform there the goal task with the needed positional accuracy.
- basic **human perception capabilities** unconsciously used by the operator when working on the spot, must be reproduced in some way and reported back to the remotely located operator.

The first capability is essential if automatic features are implemented to aid the operator work (collision avoidance, automatic grasping, specified trajectory following, etc.) or when on-line model-based graphical simulation is to be offered. The second is needed to allow easy and efficient manual remote control (stereo-vision, force feed-back, enhanced camera images, etc.).

2.2 Type of measurements

The basic requirements can be listed as follows:

- Knowledge of the **absolute position** of the manipulator with respect to its base, i.e. position and orientation of the end-effector, and position of all links. Both the best estimate value, and the estimated error, are to be known.
- Knowledge of the **relative position** of specific parts of the manipulator with respect to its surroundings, in particular the end-effector. This knowledge is required to cope with uncertainties related to the absolute position (e.g. deformation under load) and to the environment (object position uncertainty, unknown obstacles).
- Perception of the **contact characteristics** when the manipulator touches or grasps an object (localisation of the contact, forces and moments involved)
- Perception of the **object or environmental parameters** (temperature, surface finishing, sound, vibrations, radiation level, atmosphere)

2.3 Needed sensors

The needed sensors for these different perception capabilities are the following:

- Absolute position w.r.t. references
  a. **Joints position** sensors (angular or linear depending on the manipulator cinematics) to build up sequentially the position of each link with respect to the manipulator common base.

Here, the different errors introduced at each joint will be added up to the end-effector itself. Typical sensors used are here angular optical encoders, resolvers, linear optical encoders, LVDT's, potentiometers, inclinometers, etc.

b. **Absolute position measurement** of a given part of the manipulator (e.g. the end-effector) with respect to a given reference. Typical sensors are based on optical triangulation (laser beam, 3D image analysis, etc), using as reference, either specific features of the environments (walls, edges), passive or active beacons, or the location of the detecting cameras.

- Relative position w.r.t. surrounding objects

  a. **Sensors**
  Relative position between the manipulator (mainly at the end-effector level) are usually measured by non-contact sensors using electromagnetic, optical or ultrasonic principles.
  Electromagnetic type sensors are inductive or capacitive transducers, mainly used for short range sensing at end-effector level. Ultrasonic type sensors are used for collision avoidance in mobile platform control, and for longer range measurements. Optical type sensors are based on reflecting light, triangulation systems or time-of-flight measurement. They are mainly used at end-effector level, but can also be applied for collision avoidance purposes.

  b. **Stereo vision**
  Another type of sensor, namely video cameras, can be used both in automatic control mode and in manual teleoperation mode. In order to give geometrical position information in a typical 3D situation, they must provide range information in a given way. Stereo cameras are used to give such information to the operator in order to help him in perceiving depth. However, range data can also be obtained by proper image processing of stereo vision (two cameras, multiple view angles, laser scanning for range images, etc).

- Contact perception

  a. **Force and moment** measurements are usually performed by strain-gage systems (e.g. 6-axes wrist) or by measuring the deformation of a compliant flexible attachment (using then for instance proximity sensors)

  b. **Tactile sensors** to measure contact force and location. Tactile arrays allow also to perceive the contact 2D surface shape. These sensors are essentially used for in-gripper location.

These measurements can be used to reproduce at the man-machine interface the contact perception (force feedback) or to control an automatic task (force driven tasks, assess a grasping operation).

- Object and environmental parameters
  Classical sensors, used to measure temperature for instance, can be integrated in the sensing capability of the manipulator to enhance the knowledge of the handled object, as well as to inform the remotely placed operator on environmental parameters. Other sensors could be: accelerometers, radiation monitors, gas analysers, microphones, etc.

The present paper will focus the discussion on the first two types of measurement in more details.

2.4 Accuracy

The specifications on measurement accuracy depends greatly on the type of task to be performed. Only examplative values will therefore be given here.

- Knowledge of the absolute position of the manipulator in a cell to update a CAD model and a graphical scene simulation programme.
  An accuracy of around 1 cm will be typically sufficient, as this measurement will be used for collision avoidance purpose and to help the operator in perceiving its position through simulation.
- Knowledge of the relative position of the end-effector prior to start an automatic grasping or insertion task.
  Here a much higher accuracy is needed, typically better than 1mm. However, many of these delicate tasks will have to rely more on compliant tools than on an accurate trajectory planning.
- Knowledge of the continuous relative position of the end-effector to follow a specified relative trajectory (monitoring, welding, cutting tasks, etc.).
  The typical accuracy is for some cases to be of the order of some tenths of mm.
- Knowledge of the relative position of a mobile platform for trajectory planning.
  The level of accuracy can be much relaxed here, as an error of a few cm is usually still acceptable.

2.5 Repeatability

Besides these basic accuracy requirements, however, more stringent specifications are usually be imposed on reproducibility, as a known reproducible inaccuracy can be coped with in the control system. On the contrary, a sensor degradation process that would increase the systematic error after a given time period could impair the proper functioning and reliability of the telemanipulator.

As a general rule, one could state that the reproducibility must be kept better than 1%. Proper calibration, or recalibration, procedure can usually guarantee such a value, but a good knowledge of the possible degradation or aging of the transducers is essential in assessing the measurement data confidence.

2.6 Weight and volume

Other requirements concern the transducer weight and volume. For static sensors, not located on the manipulator itself, as for instance environmental surveillance cameras, or fixed position sensors, no severe limitations can be expected, although space is usually a critical resource in a cell type environment.
For on-board sensors, and more particularly, for end-effector sensors, sharp limitations on weight and volume exist. Typical sensors should not go beyond 10 cm3 and 500 g at end-effector location, 1 dm3 and 2 kg at other places of the manipulator.
Moreover, the transducer would not protrude too far outside the manipulator envelope, without causing collision hazards. One would limit this protuberance to around 10% of the envelope dimension.

2.7 Range

The needed measurement range depends on the type of measurement. Basically absolute position sensing has a greater range, a large as the maximum reach itself (a few meters in the case of a cell work, up to several tens of meters in case of mobile platforms).
Relative position sensing for collision avoidance ranges from a few cm (collision protection sensors at manipulator level) to 1 or 2 m (collision avoidance system for mobile platforms).
Accurate position measurement at end-effector level has a smaller range requirement, of the order of a few mm to some dm.

2.8 Aperture or field-of-view

The aperture of the sensors determines the area observed by one particular transducer. A large aperture is acceptable for anti-collision sensors, as this will limit the number of sensors (and the cables bundle) needed. However, very small aperture, and therefore low side effect sensitivity, is required for distance measurement. Proper sensor signal processing can in some case overcome the aperture inaccuracy (phased array disposition, use of sequential data in case of moving transducer, etc).

2.9 Response time or real-time capability

Position sensors used for off-line measurement (start reference, object recognition, etc) may need some process time delay without impairing the telemanipulator efficiency. The main limitation is here related to the man-machine interface and the wait time perceived by the operator.

On-line measurements (collision avoidance systems, path following tasks) require a small measurement loop time. However, the required response depends on the manipulator speed. As most of the tasks performed in tele-operator mode (semi-autonomous) are low speed ones, response times of a few 10ms are usually acceptable.

2.10 Target material influence or detection capability

Under typical situation, all sensors should be at least sensitive to metallic targets (stainless steel). In some circumstances, other surface materials should also be sensed: mild steel, concrete walls, plastic waste drums, etc. A particular constraint can appear from plastic gaiters placed around some parts of the manipulator or around objects to be handled or avoided. These soft plastic covers, widely used in the nuclear applications to avoid contamination, can represent specific challenges in some autonomous task sensor-based control.

2.11 Power consumption

For autonomous mobile platform, limited power resources are provided by on-board supplies. The power consumption of all sensing devices should therefore be minimal in this case. This limitation is not so critical for gantry crane applications, or when an umbilical link is considered.

2.12 Type of output and interface requirements

The main requirements related to signal processing and interface to the control system are typical of any instrumentation network (1). These requirements concern

- the cabling (mounting, shielding, grounding, testing)
- the signal processing (ADC, amplification, correlation, integration)
- multiplexing and coding (analogue, digital)
- interface to computer system or control board (bus standards)
- high reliability by proper error detection and signal validation (in case of transducer failure)

However, specific requirements are related to the usual constraints in nuclear robotics to have the least possible electronics on board, and to get therefore long cable lengths between transducer and processing electronics. The problems that arise are then

- to insure proper noise-free transmission without pre-amplification and to compensate the possible signal loss or capacitive cable effect.
- to reduce the number of cables by multiplexing systems. A large cable bundle gives particular difficulty in mechanical handling.
- to envisage the possibility of wireless transmission (electromagnetic or optical) in some installations.

A concise and easy way to interpret the data is also of primary importance. Too much data need long processing time and get sometimes ambiguous results due to the complexity of the interpretation software. Data should sometimes be provided only when needed, when action is required. Background preprocessing, as for instance level or rank detection, should be envisaged.

3. Nuclear environment requirements

The nuclear environment is characterised mainly by the presence of ionisation radiation and contamination that can impair the proper functioning of sensors and their related processing hardware. The radiation levels are however quite different from one application to the other (2)(3).
A tentative categorisation is given here. It is evident that particular tasks can correspond sequentially to different categories, when for instance the manipulator is handling a highly radioactive object (a fuel element for instance) in an environment where the rest of the work occurs at much lower dose rates. For design purposes, the highest constraints are naturally to be taken into account, but with the work duration in mind.

Four categories could be considered (Fig. 2):

A. Extremely radioactive environment

This is specific to in-core maintenance operation, post-accidental intervention, reprocessing plants (parts in direct contact with fuel), etc). Human intervention is not possible. Remote operation is therefore an absolute necessity.
The dose rates are to be expected between 0.1 and 10 Mrad/h. Mission times are however usually relatively short (100 to 1000h), so that an integrated dose of 1 to 10 Grad is typical. Replacement of sensitive parts must be considered between missions.

B. Highly radioactive environment

This is specific of all operations around the reactor vessel, including dismantling work, as well as typical environmental condition in hot cells.
The dose rate to be expected here are around 0.1 to 10 krad/h, with an integrated dose of around 1 to 100 Mrad (mission time of the order of 1 year).

C. Highly contaminated environment

This is specific to most cell work, as well as to first stages of decontamination procedures. many dismantling operations are also in this category. Human operation is sometimes possible using particular protection, but with high dose cost (therefore high personnel cost and long delays). Remote operation brings shorter delays, enhanced radioprotection optimisation, and lower overall costs.
The dose rates are to be expected from a few rad/h to 500 rad/h. An integrated dose of about 100 to 1000 krad is to be foreseen.

D. Low radiation environment

In most operations inside the reactor building (red zones), such as inspection, monitoring, light maintenance work or decontamination processes, human intervention is usual. However, due to the reduced dose allowance, radioprotection optimisation (ALARA) and human error minimisation, telerobotics is more and more considered as an interesting alternative. The dose rates are here less than a few rad/h, with an integrated dose of about 1000 rad.

For the environmental tolerance of sensitive components placed on the manipulator, category D will not impose particular constraints, as usual reliable components will not suffer from these low radiation levels. Any robotic feature can be used, as sensors, cameras, etc, and even on-board computers can be envisaged.
Category C requires more specific design and careful choice of components, but usually standard products can still be used, let it be with stricter specifications. Electronics and processing boards can be used, but with so-called rad-hard (but commercially available) components.
Category B is more difficult, as no complex electronics can be used. Sensor materials and cable insulation requires particular attention and the choice of components is limited. Simple electronics can be envisaged, but proper design must be made to stay tolerant to individual component degradation.
The toughest category A excludes any electronics, and limits the choice to purely passive sensors with carefully chosen materials. Cable links are also to be properly specified.

The nuclear environment does not impose only requirements about radiation tolerance. Some additional specifications concern the following aspects:

- **contamination**
  The manipulator and its related equipment will be subject to contamination risk, and must therefore be easily decontaminated. This requires simple external surfaces, with thigh joints, or protective plastic envelopes.

- **repairability and replacement capability**
  Any repair in a nuclear environment requires either the dismounting and decontamination of the failed component, or the presence of operators inside the radioactive zone. Particular attention must therefore be put on easy dismounting and replacement of sensitive components. This is even sometimes the primary requirement, above any radiation tolerance, as the option of "throw away" sensors or even cameras, is in some cases the cheapest and most reliable alternative.
- **reliability**
  Consequences of failure can be very serious in nuclear environment, with high cost, delays and man doses involved. The requirements on reliability of any system are therefore much higher than in corresponding industrial applications of robotised machines. For the particular case of sensing, validation processes of measured data, redundancy of sensors, and multi-modal capability or cross correlation of information from different origin is essential, before specific tasks are launched. Such features increase the confidence level of the provided data. This is also the case for human error avoidance by automatic watch-dog features, such as collision avoidance, automatic check lists, etc, where autonomous sensing plays a leading role.
- **mixed control mode**
  Due to the complexity of the environment and the level of reliability required, most remote operations in the nuclear field are and will be characterised by a so-called mixed mode operation, or semi-autonomous control, where both man and computer shares the control burden. This will impose some specific requirements on the sensors and their related data processing, as their interpretability by a human operator is central. Their use in the man-machine interface must always be taken into account, and not only their integration in an automatic control loop.

4. Review of existing techniques for internal position sensing

In order to control any manipulator, the first requirement is to get a knowledge of the position of all its links w.r.t. each other, and with reference to a fixed coordinate system, usually the manipulator base or first link attachment. This requires that any joint (angular or linear) be measured in position. The sensors used for this purpose are sometimes called internal sensors, as they are part of the manipulator itself.

Most machines are equipped with one of the following systems: potentiometers, linear (or rotary) variable differential transformers (LVDTs), resolvers, encoders and inclinometers. These systems will be presented and discussed hereafter.

4.1 Potentiometers

The potentiometer is the earliest form of displacement transducer to be used in control applications. It was widely used in first generation robots. In its standard design, the device consists of a uniform coil of wire or a film of highly resistive material such as carbon, platinum or conductive plastics, whose resistance is proportional to its length. When a regulated DC source is applied across the resistance, the output signal is the voltage drop between a movable contact (wiper) sliding on the element and one of its terminals. Potentiometers can be applied for linear or angular displacements. This approach has many drawbacks related to mechanical loading due to friction and inertia, limited bandwidth, electrical loading errors and non-linearity as well as degradation of the contact due to wear and environmental influences. Although it offers a low cost and versatile solutions in some applications, care must be taken with this type of transducer for applications where long term reliability and repeatability is needed. However, when periodic maintenance is foreseen, and easy replacement possible, potentiometers can offer competitive solutions.

4.2 LVDTs

Linear (and rotary) variable differential transformers are commonly used for displacement measurements. They overcome most of the shortcomings of the potentiometers. Based on the mutual-inductance principle, they consist of a primary coil wound on a non-magnetic housing, and a symmetrically placed secondary coil (eventually two symmetrically places secondary coils). A ferromagnetic core allow for magnetic coupling between them. The primary coil is energised by an alternating current. This generated an induced voltage in the secondary parts. The axial (or angular) position of the coupling core modifies the reluctance of the flux path and determines the level of flux linkage. By connecting the armature at one part of the joint and the coupling core to the other, a secondary voltage can be measured that is directly proportional to the relative displacement.
Linearisation can be obtained by proper signal processing. These systems have very good resolution, high accuracy and can have long ranges. They do not establish direct mechanical contact between the two parts of the joint, avoiding mechanical loading and wear. They also provide directional indication. However, they require fine mechanical design to guarantee exactly reproducible movements (colinear for translatory sensors, e.g.).

4.3 Resolvers

Resolvers are mutual induction transducers commonly used to measure joint angles in industrial manipulators. It consists in a rotor made of a single two-pole winding for instance, activated by an AC supply through a slip-ring brush mechanism, and a stator with two sets of windings placed 90 degrees apart. The induced voltages in the two stator windings are orthogonal to each other, depends on the geometric and material characteristics of the system and have a fluctuation with a period proportional to the angular position between rotor and stator. As the sensor delivers the wanted data in the form of two orthogonal trigonometric functions, it is also used to feed directly dynamic control algorithms for the joint torques, saving thus processing time. An alternative disposition is to feed the two stators by a pair of orthogonal AC voltages, and to measure the phase of the induced voltage on the rotor. This last system avoids the need of slip-rings and brushes.

The advantages of resolvers is to provide fine resolution and high accuracy with a very robust construction. However, the information is non-linear, bandwidth is limited by the supply frequency (this frequency has to be at least ten times the maximum frequency of interest for the measurement), and mechanical brush contacts are needed in some designs.

It is to be noted that dedicated electronic circuit are presently available to linearise the resolvers signals and to code the output in a digital format, easing thus their interconnection to a digital input/output control system. This processing electronics can be placed remotely from the resolver transducer itself.

4.4 Encoders

Shaft encoders, both incremental or absolute, come in many types, but optical models are the most common on robotic equipment. It is directly mounted on the joint to reduce any coupling errors (backlash, shaft flexibility, loose fit, resonance, eccentricity, etc). They are used both for angular and rectilinear movements.

4.4.1 Incremental encoders

Optical incremental encoders are based on an encoded disc where a pattern of equally spaced alternating transparent and opaque areas is imprinted as a circular track. One side of the disc is illuminated by a light source, such as a light emitting diode. On the other side, two light sensitive probes (phototransistors, e.g.) are for instance located along the track at half window width offset.

As the disc rotates, two pulse trains are generated as output. Direction of movement is derived from the phase shift between them. Counting the pulses allow one to keep track of the movement.

However only relative position increments are measured, and an absolute reference must be available at the beginning or at given times to reset the system. Information on movement speed can also be obtained by proper timing of the output pulses. The digital characteristics of the sensor output makes it easy to interface to a manipulator controller. Moreover, optical incremental encoders are quite robust and cost effective, with a high level of accuracy if simple movements are present. Resolution of the order of 0.01 degrees are achievable (4). These sensors are however sensitive to vibrations, or mechanical oscillations, and their main drawback is their lack of any absolute reference, allowing therefore in some instances to build up errors without assessment capabilities.

4.4.2 Absolute encoders

Absolute optical encoders are physically built in a way similar to the incremental types, but they generate directly coded data. No pulse counting or timing is involved. Several tracks are imprinted, providing a binary code for each position. A bank of detectors records the unique code for each position. The binary word obtained can be directly outputted to the controller. The accuracy of the output is dependent on the number of tracks, corresponding to a bit length, and depends critically on the proper reading of the code at each position (5). Any deformation of degradation of the coded disc results in unreliable and ambiguous data.
Absolute encoders have the main advantage to provide instantaneous information, without needing a continuous measurement. They are however more complex, both in mechanical and electronic design.

4.5 Inclinometers

Alternative information on the position of the different links of a manipulator can also be obtained by measuring their inclination w.r.t. the vertical direction. Tilt sensors usually features a pendulum of some sort suspended from the shaft of rotary displacement transducer. Rotation of the body will give rise to an output change in the measured signal. Other design consist in a modified accelerometer where the component of gravity parallel with their axis of sensitivity is recorded. In this last case, proper design can eliminate in great part the sensitivity to vibration, as they are actually used, at least at low levels, to perform the measurement. However, for both types, dynamic inertial effects impede their use on line. They can on the other hand be of high use to back-up angular measurement systems or to provide reference situation to incremental transducers.

5. Review of existing techniques for external non-contacting position sensing

Sensors used to perform a positional measurement without contact can be categorised depending on the portion of the energy spectrum they cover (acoustical, electromagnetic, optical), or depending on the physical principle used to derive the positional data (intensity reflection, triangulation, time of flight, phase or frequency modulation, interferometry, etc.).

5.1 Intensity reflection

Techniques involving return signal intensity determine the distance to an object based on the amplitude of energy reflected from the object's surface. Such a principle is mostly used with light for short ranges. The infrared region is usually chosen, due to the common availability of sources, and the relative independence on the ambient illumination. Pulse triggered measurements are also preferred to lower the influence of environment fluctuations.

The sensor is composed of an emitter (often a light emitting diode) and a receiver (e.g. a phototransistor). An appropriate optics focuses the light beam. Different types can be envisaged:

5.1.1 Break-beam type

The break-beam type senses the target when it breaks the optical beam between emitter and receiver. Such a type can only be used for very specific applications in nuclear robotics applications (protecting interdicted zones, object grasping assessment).

5.1.2 Beacon reflection type

The beacon reflection type emits light towards a mirror (beacon) reflecting the signal towards the receiver colocated with the emitter. The object is detected, either when it breaks the beam, or the retro-reflector is placed on the object itself. Applications are similar to the break-beam type. The location of retro-reflective beacons on objects is usually quite unrealistic in nuclear environment, although reference points (on the cell walls for instance) can be so highlighted for easy recognition and position detection). Beacon can eventually be more elaborate and contain a coded information, as for instance a bar code.

5.1.3 Diffuse type

With the diffuse type, only the light diffused back from the object itself (without cooperative target reflector) is measured by a receiver colocated to the emitter. Here near-infrared modulated light is always used to reduce the influence of ambient light sources and to get a sufficient signal to noise ratio. Some realisations add a convergence feature where emitter and transmitter are angled w.r.t. each other so that detection occurs only over a small portion of the range.
Targets outside this range are not detected. This feature decouples the sensing technique from any dependence on the reflectivity of the surface.
This principle has the advantage to be very simple in its application and very cheap due to the easy availability of components. However, it mainly suffers from its great dependance on the surface reflectivity of the target object and from its very short range. For very known surfaces (as in production lines), the technique can be used with a high level of reliability. In more general applications, and in particular in non-structured surroundings typical of nuclear industry, it is limited to very short range proximity switching (collision avoidance), or to measure sharp contrasts in return signal (for edge detection for instance, or material/texture discontinuities). The sensor must however then be moved w.r.t. the target object.
Another application of diffuse type sensors is the use of two identical point source LEDs positioned a known distance apart, with their incident light focused on the target surface. The emitters are individually triggered and the returned energy is sequentially measured by the detector. If the two LED were located at the same distance from the target, the two measurement should be identical. Any difference can be related to the fact that one source is closer to the surface. This can be used to measured absolute distances, and to cancel out varying surface reflectivities (6).

5.2 Triangulation

Ranging by triangulation has been used for ship navigation since very long times. It is based on the simple property that a triangle is completely determined when its basis and the two adjacent angles are known. This principle can be applied either passively (two detectors) or actively (one source, one detector). Usually the signal used is light, but some applications use ultrasonic waves also (7)(8).

Ranging can be applied for two categories of applications: the so-called "eye in the hand" type where 3D perception is obtained at the end-effector level for the scene in the immediate vicinity of the manipulator hand, and the longer range open view mode more directed towards mobile platform navigation applications.

It is mainly in the former area that many proposals have been formulated and applied. This is mainly due to the sparseness of data required and to the usual better definition of the scene to analyse, reducing thus the processing complexity to real-time capabilities (8)(9).

5.2.1 Stereo vision

One application of passive triangulation is the so-called stereo vision, analogue to the human eyes depth measuring capability. A 3D object is viewed from two locations on a plane normal to the direction of vision. The received images are shifted laterally with respect to each other.
This displacement is inversely proportional to the distance of the object. Practical ranging systems use two cameras, possibly collocated, or one moving camera (10). The range measurement occurs in different steps: a reference point (or object) must be recognised on one image, the same point must be located on the second image, their relative positions in the images must be measured, and finally the distance can be deduced. This technique suffers from two main drawbacks: it is quite costly and cumbersome, as cameras must be used, and the processing is very complex, as the recognition of reference points are quite difficult in random unstructured scenes. The technique is therefore only used in more global applications, where not only a distance is measured, but where more complex object recognition and localisation is performed, let it be on a small set of known targets. Stereo vision is on the other hand an important part of the man-machine interface, to provide the operator with a depth view, but by using then the human image processing capability (11).

5.2.2 Laser triangulation

Laser triangulation is an active application of the triangulation principle, where a given illumination of the scene is provided by a laser source. A camera, that can at the limit be only a liner detector, is offset from the source by a known distance, and oriented to look to the illuminated spot. The range is basically determined from the known baseline distance (source to detector) and the position of the laser spot image in the measured pixel array of the camera (6)(12).
Different alternatives have been designed:

a. The laser beam can be **structured** as a plane (line illumination) or a grid pattern. The advantage is here that the range can be determined for many points of the scene at the same time, increasing therefore the 3D scene perception, but keeping the processing time in reasonable limits, compared to complete stereo vision processing (13).

b. A single element detector (a photodiode e.g.) is mechanically **scanned**. the range is then determined from the scan angle at the moment where the observation is taken.
c. A **complete 3D scene** can be measured (range array) by scanning either the illumination beam in two directions, or a plane illumination in one direction (14). This last possibility reduces the dead time provided by the (mechanical or optical) scanning.

The triangulation principle suffers naturally from the two known drawbacks: **occlusion** problems when the complete object surface cannot be illuminated or seen from one particular angle due to the presence of obstacles or due to inherent portion of the object itself, and **accuracy** limitations due to angle measurement difficulty and basis length limitations, especially for eye-in-the-hand applications.
However, this technique offers important advantages: it provides the control system with a lot **more information** than just a distance (complete scene 3D pixel matrix, object shapes, references recognition, etc.), and for a quite **modest processing time**, as the image to analyse is much more structured and sparse, than usual camera images.

5.2.3 Other 3D vision techniques

Besides the most common stereo cameras and triangulation sensors, other techniques can be used for depth perception.

a. The **stadimeter** (15) as used in navigation for nautical applications measures the distance of objects of known size by measuring the angle of vision or the height of the recorded image. This is only possible when a recognition of the target can be performed before measurement. An automation of the process is difficult to achieve in a purely unstructured environment, at least as long as known beacons cannot be placed at specified locations. Such an installation is usually excluded in many nuclear typical situations.
b. The **swept focus** technique (6) uses a modified video camera with a single lens of very short depth of field to produce an image in which only a narrow interval of range is in focus at a given time. The lens (or the detector plane) can then be positioned sequentially to different locations in order to view different range slices. Here also, as in the previously presented technique, the interpretation requires a prior recognition of the object or of some reference features. When this reference feature is in focus, its distance can be derived from the lens position.

5.3 Time of flight

On the contrary to the previous ranging approach where triangulation is used in one way or the other, typical radar approach obtains distance information by interrogating a scene with only one beam of energy wave. This can be performed by electromagnetic, light or acoustic signal. It can be pulsed or modulated. But in all cases, the beam is sent towards the target, reflects back to the emitter/receiver collocated system and is there detected. Distance information is obtained by determining the time it took to flight to and back, either by direct timing of a pulse, or by measuring the phase shift between sent and received signal (amplitude or frequency modulated).

5.3.1 Pulse modulation

In this type of technique, the time shift between outgoing and return pulses is used to measure the distance. The most popular sensor based on this approach uses acoustic wave. It is the **ultrasonic ranging system**, mounted on many robotic applications.
Typical system can measure distance from a few cm to several meters, with an accuracy better than 1 %, but are quite sensitive to environmental parameters of the atmosphere (temperature, pressure, turbulences, etc). Also specular reflections can greatly impair the measurements by giving completely discordant data, and broad aperture (from 10 to 30 degrees) reduces the interpretability of the results. They are mainly used for collision avoidance, but find interesting uses also for path following and end-effector approach movements. Many improvements have been studied, both to improve the data assessment (16) or to allow for electronic beam scanning (17).
Applications with electromagnetic waves **(RADAR)** and light **(LIDAR)** are only limited to longer ranges (greater than several meters at least) due to the high travel speed of these waves in air. Subnanosecond timing circuitry is needed. In practice, the accuracy decreases sharply for smaller distances. However, they have much smaller apertures and are well suited for navigation or tracking purposes, but their application at end effector level is not feasible.

5.3.2 Phase and Frequency Modulation

Time of flight approach with light or high frequency electromagnetic waves can be better suited for shorter range applications when no pulses are generated, but instead continuous waves with a suitable modulation (phase or frequency).

a. Phase shift

When a continuous wave energy beam is aimed at a target, the return signal experiences a shift in phase. The relative phase shift is measured by comparing a portion of the original beam and the returning one. The phase shift is a function of the round trip distance the wave has traveled. The system can also be applied with an amplitude modulation of the continuous wave pattern.
This technique involves usually a complex and time consuming signal processing, as accuracy can only be obtained by integrating several successive measurements (18). Real time applications are therefore difficult to achieve.Moreover, several measurements at different modulation frequencies are necessary to remove inherent ambiguities. It is therefore less useful for continuous control at end effector level, but finds interesting applications as navigation aid in open field mobile platform movements.

b. Frequency modulation

Frequency modulation is widely used in radar altimetry for long range applications. A continuous wave electromagnetic or laser beam is modulated by a periodic triangular signal, varying the frequency around a given mean value. This signal is reflected back from the target with the frequency curve displaced along the time axis by an amount equal to the time required for the wave to travel to the target and back. The beat frequency produced is proportional to the distance to be sensed. Typical accuracies obtained with laser light at a distance of 1 m are of the order of 0.1 % (19).
This technique has the advantage to require only one measurement to get an unambiguous measurement of the distance. However, the accuracy is a very critical function of the stability of both laser and frequency ramp. Its application is therefore doubtful at this stage for end effector applications.

5.4 Interferometry

Interferometry is a very accurate measurement technique for distance sensing. It provides however only relative data (6)(12). This ranging technique is based on the resulting interference patterns which occur when two light waves caused to travel different paths are compared. If the length of one of the two optical paths is changed, the two beams will interact in such a way that clearly visible constructive and destructive interference fringes are produced. These fringes are counted, and the relative distance displacement can be deduced. As no absolute values are obtained, a reference point must be set in the beginning and subsequent data will be distances from this reference. The beam is assumed never to be broken or the target is lost.

Due to the high complexity of the processing as well as the very high stability needed for the laser light, only few applications of this technique have found their way outside the laboratory environment. However, it is well suited for tracking systems, where the position of the end effector of a telemanipulator is measured from a fixed reference point (20). Nuclear applications add anyway difficult constraints, as turbulence free atmosphere, continuous direct line of sight between the fixed tracking system and the end effector, presence of a retro-reflecting surface at the sensed point, etc.

5.5 Electromagnetic proximetry

Electromagnetic proximity sensors have been developed to gain position information in the close-in region (in the order of a few cm), extending the sensing range beyond that afforded by direct contact tactile sensors. By careful signal processing, very high accuracy have been achieved. A very broad field of industrial applications lie also in the proximity switching, where non-contact approaches have many times replaced the traditional mechanical limit switches. Two basic types can be considered for robotic applications, the inductive and the capacitive ones. A third type, based on permanent magnet or Hall effects, is confined to particular situations.

5.5.1 Inductive proximity sensor

An inductive sensor is sensitive to the presence of a conductive (e.g. ferrous or non-ferrous metals) object placed in its vicinity, through its influence on an HF oscillator circuit located in the sensing head (frequency from 10 to 3000 kHz). In the so-called Eddy Current Killed Oscillator design, the HF oscillator generates an oscillating electromagnetic field around an inductive spool. Eddy currents are induced in any metallic objects placed nearby. HF energy is absorbed, and the oscillator works to reduce these currents by drawing more current from its power source. This current changes in proportion with the distance of the object to the sensing face of the spool. These sensors have a range extending typically up to 10 cm depending on their size. Measurement is performed continuously, with response time only limited by the related processing electronics. Very robust construction can be obtained, providing high resistance to harsh environment (vibration, temperature, corrosive atmosphere, and also ionising radiation). The main drawbacks are the limited ranges, the sensitivity to material properties and target shape.

5.5.2 Capacitive proximity sensor

Capacitive sensors are similar to inductive ones, but use in the sensing head a capacitive circuit, sensitive to any metallic or dielectric object placed in the vicinity. Such system is therefore very interesting to sense non-metallic targets (plastics, concrete, liquids, etc.). Usually the transducer is an open capacitor connected to an oscillating bridge. A target closes the magnetic path and changes the capacitance of the transducer. this change varies as a function of the distance from transducer to target, but also of the target properties, shape and approach angle. The sensor can therefore be used for presence detection, but range determination is only possible using a prior knowledge of the target itself (material, orientation). Advances in electronic technology have improved these sensors, providing very high level of accuracy. However, they suffer from the same limitations as the inductive sensors, w.r.t. range.

A slight modification of the same principle is used in capacitive collision avoidance detectors, where a ungrounded (floating) metallic structure is used as an antenna. The capacitive part of the antenna impedance is influenced by any grounded dielectric or metallic object coming in its surroundings. No real range determination can be obtained, but this type of sensor is very interesting due to the flexibility of its coverage (the metallic antenna can be configured to sense any specified volume), the great volume covered using only one sensor (and therefore one cable), and the purely metallic characteristics of the transducer (making it quite suitable for any hazardous environment).

5.5.3 Permanent magnet sensors

These sensors are usually built around the Hall principle. When a magnetic field is applied perpendicular to a fixed current flowing through a semiconductor material, a voltage is produced that is proportional to the flux density. This flux is changes by the movement of a permanent magnet. This magnet can be part of the target, of integrated into the transducer. In this latter case, only ferrous targets are sensed. When unstructured environment are considered around a robotic application, as it is usually in all nuclear applications, this type of sensor cannot be easily applied. Only particular situations, similar to industrial automation setting, can lead to the installation of such a technique. It must be taken into account that sensing range is here very short, but that high reliability and accuracy can be achieved.

6. Discussion on nuclear adequacy and short list of suitable techniques

The different techniques that have been presented in the previous chapter have all found some concrete application in industrial automation and sensor based robotics. They are however not all of them so adequate to be implemented in a nuclear environment, taking into account both the environmental conditions (radiation mainly) and the unstructured aspects of the surrounding objects. This chapter will discuss them in view of their nuclear application, selecting the most suitable ones, and stating for each of them the needed improvements.

This list confines itself to the internal sensors of the manipulators, to the sensors used at the end-effector level to measure its position w.r.t. the objects to handle, and to the detectors needed to protect the manipulator links of possible collision. Cameras and long range navigation sensors are not considered.

6.1 Internal position sensing

A. Resolvers

General description : paragraph 4.3

Typical use :
a. Angular measurement of manipulator joints
b. Angular measurement of auxiliary motors (x-y-tilt tables, crane position, mobile platform wheels, etc.)

Advantages :
a. Robust construction
b. Available signal processing for digitalisation

Drawbacks :
a. Non-negligible size
b. No direct usable coded output

Radiation tolerance :
As the resolver transducer consists only in spools wounded on metallic armatures, the main sensitive parts are related to electrical insulation (connecting wires, spool wires, connectors) and to the resin filling of the windings. Proper choice of material is to be made, the resin problem being the most critical for high radiation environment (for total dose higher than 10 Mrad typically). Tests have been performed at CEN/SCK on specially constructed resolvers. A total dose of more than 100 Mrad could be achieved before the first degradation appeared (21).

B. Encoders

General description : paragraph 4.4

Typical use :
a. Angular or linear measurement of manipulator joints positions
b. Similar measurement on related auxiliary equipment or tools

Advantages :
a. Digital data directly interfaceable and standard for most control systems
b. Compact and small-weight sensors, with high accuracy
c. Absolute measurement capabilities

Drawbacks :
a. More delicate construction, sensitive to vibration and shocks
b. Both optical and electric components
c. Only incremental data in the simple versions

Radiation tolerance :
Typical industrial encoders are not radiation resistant, mainly due to the sensitivity of optical components. Two possibilities exist to increase the tolerance of such devices to ionising radiation:
a. to design the optical link (LED and detector) with radiation tolerant components, that are available to some extent, but still need some testing, if high doses are considered. Tests have been reported on a hardening up to 100 Mrad (22).
b. to remove the sensitive components outside the radiation field by connecting the encoder through optical fibres. Such encoders are commercially available, but custom-made design is needed with the proper choice of rad-hard fibres (23).

C. Potentiometers

General description : paragraph 4.1

Typical use :
a. Angular measurement of manipulator joints
b. Angular measurement of auxiliary motors (x-y-tilt tables, crane position, mobile platform wheels, etc.)

Advantages :
a. Simple construction
b. Low cost solution

Drawbacks :
a. Reliability for long term use
b. No direct usable coded output

Radiation tolerance :
The simple construction of potentiometer transducers make them quite resistant to moderate doses of gamma radiation. The main source of degradation is related to electrical insulation. In most cases, wear-off type degradation processes will be more significant. Decalibration will be the usual appearing phenomenon, making it actually perverse if no detection capabilities is possible. Potentiometers are mounted on some nuclear telemanipulators. However, they are regularly replaced at maintenance turns, usually after a total dose lower than 1 Mrad.

D. LVDT

General description : paragraph 4.2

Typical use :
a. Displacement measurement of linear links on the manipulator or on auxiliary translators.
b. Measurement of the deformation of the manipulator structure in case of high payload or vibrations, eventually in order to measure efforts imposed on the end-effector.

Advantages :
a. Well established technique with very robust realisation
b. Very high accuracy possible

Drawbacks :
a. Contact needed on the target object, with the related forces involved.
b. Non-negligible sensor length.

Radiation tolerance :
LVDTs are mainly constructed with metallic armatures, and wire windings. As for resolvers, attention must be put on insulators and windings resins, but taking into account that only very small currents are involved here. By proper choice of these materials, high radiation tolerance can be obtained (up to 1 Grad).

E. Inclinometers

General description : paragraph 4.5

Typical use :
a. Angular position of the different links of the manipulator as a redundant measurement for reference position control.
b. Angular position of auxiliary equipments, as for instance tilt tables.

Advantages :
a. Small sensors with no direct interface with the drive and gear mechanisms
b. Absolute measurements w.r.t. verticality (in still-stand positions)

Drawbacks :
a. Problems due to inertial and vibrational behaviour
b. Limited high accuracy angular range

Radiation tolerance :
Very few data are available on radiation tolerance of inclinometers. However, accelerometer types should have tolerances similar to the ones valid for accelerometers. Radiation hardening has been achieved to high levels (up to 1 Mrad) provided careful design is guaranteed. Higher tolerance should be assessed. Pendulum type sensors can experience degradation due to movement jamming by lubricant thickening, or contact resistance increase. Proper evaluation of existing design is still to be performed, both theoretically and experimentally.

6.2 External non-contacting position sensing

This list provides different sensor types that offer the best potentialities in nuclear applications, taking into account the specific requirements of such an environment. For instance, the following types are not cited, because they have main drawbacks in their present state of the art. However, due to rapid technological advances, some of them could in the future be considered, provided some improvements could be realised on their specific problems:

- **stereo vision** : Although such systems can offer very helpful support to the operator, as part of an efficient man-machine interface, their use in an automatic control mode must be limited to the recognition and localisation of a particular limited set of objects. Their general use as range perception device in unstructured environment is mainly limited by the long image processing time involved and the bulkiness of radiation resistant systems.
- **laser ranging systems using frequency or phase modulation** : These systems are up to now mainly adapted for long range detection, for instance in navigation tasks. They are still too voluminous to be used at end effector level, and their accuracy decreases at short distances. Rapid developments are however present in this field and the use of fibre optics to remove any electronics outside the radiation area could be determinant arguments to consider them in the future as hands-on perception devices.

- **electromagnetic radar systems** : The applications of electromagnetic signals reflection in a very clustered environment where the majority of objects are metallic gives rise to many parasitic effects. Moreover, short range applications have limited accuracy, and still involve very bulky transducers.
- **interferometry** : This principle requires a very sophisticated installation, and can only be envisaged to measure the position of a manipulator from an external stand of view. Such disposition is difficult to achieve in typical nuclear environment.
- **permanent magnet proximity sensors** : These sensors require the installation of target pieces on the objects to handle, or are limited to ferrous materials. Their application is therefore not to be envisaged generally, except in particular situations. The discussion given below for inductive sensors is in fact valid in these cases also.
- **triangulation techniques using one single camera** (stadimeter, swept focus) : These approaches require the a priori knowledge (angular localisation and recognition) of reference objects. Such a procedure needs a very heavy image processing if the reference is an existing feature of the environment. If special reference features are introduced (as for instance high contrast marks, bar codes, etc.), their general application in any typical nuclear environment is usually not feasible.

The following six types on the other hand are in such a development state that direct application in nuclear environment is feasible. However, this does not mean that specific improvements are still to be achieved, but they are eventually achievable with the present state of technology.

A. Diffuse reflective infrared sensor

General description : paragraph 5.1.3

Typical use :
a. At end-effector level, to sense the immediate surroundings of the gripper (range of the order of a few cm)
b. Along the manipulator body, at critical points (protuberances) to sense possible obstacles and avoid collision.

Advantages :
a. Small size, low cost, and easy replacement or maintenance.
b. Low influence of ambient lighting condition
c. Sensing of metallic and non-metallic objects

Drawbacks :

a. Large influence of surface reflectivity. No reliable ranging possible, except for known reproducible targets. It gives however good relative data (obstacle/no obstacle or edge/discontinuity location).
b. Short range and narrow coverage
c. Influenced by temperature of target

Radiation tolerance :
The optical parts (source, detector, lenses, optical fibres) are subject to degradation processes, leading eventually to important attenuations. Two improvement directions can be followed:

a. using rad-hard LED and phototransistors, as well as special nuclear lenses.
b. using rad-hard optical fibres (and lenses) to locate sources and detectors remotely (23).

No off-the-shelf rad-hard sensors are however commercially available.

B. Beacon reflective and break-beam infrared sensor

General description : paragraphs 5.1.1 and 5.1.2

Typical use :
These applications of the intensity reflection principle find only limited uses in typical nuclear environment, mainly due to the fact that they require some provisional installation of fixed systems in the work space. Two typical systems can be cited:

a. Measurement of the position of reference passive beacons located at specific spots in the manipulator surroundings, on target components or on walls..
b. Detection of the presence of an object in the gripper (gripping task assessment) or protection of a given work space by a light curtain.

Advantages :

a. Simple technique, with small sensors, except for beacon position measurement.
b. Low influence of ambient light conditions
c. No influence of surface reflectivity of objects

Drawbacks :

a. Necessitates some fixed installation in the work surroundings.
b. Large sensor volume for beacon position measurement (mechanical rotation and tilt systems)

Radiation tolerance :
The discussion presented for the diffuse sensors is still valid here, with the additional aspect of degradation of the reflectivity of passive beacons.

C. Laser triangulation sensor

General description : paragraph 5.2.2

Typical use :
a. measurement of frontal distance at end-effector level to target (short ranges, up to 30 cm typical).
b. 3D scene perception by combining structured light (laser plane) with manipulator movements.
c. Laser triangulation can also be used as navigation aid for mobile platform control (but with longer ranges)

Advantages :
a. Accurate ranging data, independent on target material, orientation or illumination.
b. Capability to get not only frontal distances, but a complete 3D image.
c. Image analysis less complex than with normal video images

Drawbacks :
a. Needs free optical paths between target, laser source and detector point (false data in case of shadowing).
b. Accuracy is dependent on the source/detector base length, giving therefore spatial hindrance risks, especially at end effector level.

Radiation tolerance :
Laser sources, related optics, and certainly detectors, are normally quite sensitive to radiation. Degradation will be felt as illumination power attenuation, detection efficiency and signal to noise ratio decrease. Two ways can be followed to increase the radiation tolerance:
a. Remote location of both source and detector, by transmitting light through optical fibres, that must be chosen with sufficient radiation hardening characteristics (23).
b. Choice of rad-hard components. That is in some measure feasible for laser source and optics. However, radiation hardened detectors (1D or 2D arrays or tubes must be chosen) can only be found in very high price ranges and with unpractical dimensions.

D. Ultrasonic ranging system

General description : paragraph 5.3.1

Typical use :
a. Short range measurement at end-effector level (typically 10 to 20 cm)
b. Longer range measurement to control approach movements (typically of the order of 1 m)
c. Presence detection for collision avoidance, along the manipulator links, or on the mobile platform.

Advantages :
a. Measurement independent on target material
b. Wide angle coverage possible for collision avoidance
c. Small dimension and weight of transducer

Drawbacks :
a. Specular reflection possible depending on target orientation
b. Inherent measurement delay
c. Side effects possible even for smallest possible aperture angle (10 degrees)

Radiation tolerance :
Ultrasonic transducer can be constructed with remote electronics. Therefore, only the radiation resistance of the transducer is critical. Two types of construction are possible. One is based on a capacity effect where a metallised flexible membrane generates the sonic pulses. The other is centred around a piezoelectric crystal, with a proper matching material to match the wave impedance in the crystal to the one in air. These crystals are used for material diagnostic in nuclear environment up to high doses. But here in both cases, the critical part is the material used for the flexible foil. Any hardening of this material due to radiation would decrease the efficiency or the matching. The sonic pulses would weaken out. No rad-hard systems are yet available. Research should therefore be focused on the degradation process of these matching materials, and their eventual replacement by mechanically more stable ones.

E. Inductive sensor

General description : paragraph 5.6.1

Typical use :
a. Short range distance measurement at end-effector level (typical range of a few cm)
b. Collision avoidance system for narrow passage mounting in specific applications

Advantages :
a. Very accurate data
b. Small size and weight

Drawbacks :
a. Only sensitive to metallic targets
b. Response depending on target material, orientation and shape
c. Small ranges

Radiation tolerance :
Usual industrial sensor system are usually built in an integrated manner where the needed electronics is placed in the transducer casing. Special versions exist however where the transducer consists only in an inductive spool. When carefully built with only ceramic and metallic material, such a spool can withstand high radiation doses without measurement accuracy degradation (23).

F. Capacitive sensor

General description : paragraph 5.6.2

Two different capacitive sensors can be defined: one of the integrated type, similar to the inductive sensor, the other based on the radiating characteristics of a floating metallic structure.

Typical use :
a. Applications similar to the inductive sensor, but particularly for non-metallic targets (concrete walls, protective plastic anti-contamination sleeves, plastic waste baskets, etc)
b. Collision avoidance system when the sensor is a radiating metallic structure

Advantages :
a. Collision avoidance sensor can be shaped to the area to protect, with a large coverage for only one measurement cable.
b. Integrated sensors on the other hand can be made small and very accurate for distance measurement.

Drawbacks :
a. No range measurement for radiating structure, and strict dependence of the coverage with the chosen shape
b. Signal depending on the dielectric constant of the target
c. Small ranges for the integrated sensor (a few cm).

Radiation tolerance :
The passive transducer structure offers a very high radiation tolerance, as only metal and ceramics is used, but integrated sensors containing process electronics should be avoided. Such sensors can be procured in special versions, similar to the corresponding inductive types.

7. References

(1) B. Noltingk, Instrumentation Reference Book, Butterworths, London, 1990.
(2) E. Jones, et al., Functional Requirements for Computer Assisted Teleoperators to Operate in Reprocessing Plants and Fusion Reactors, Teleman Users Special Interest Group 2, CEC Contract FI2T-0003-C, Nov. 30, 1990.
(3) A. Khairallah, et al., Functional Requirements for Computer Assisted Teleoperators in Reactor Environments - PWR Case, Teleman Users Special Interest Group 1, CEC Contract FI2T-0002-C, Jan. 18, 1991.
(4) C. Desilva, Motion Sensors in Industrial Robots, Mechanical Engineering, June 1985, pp. 40-51.
(5) D. Overcash, Selecting the Proper Position Sensor, Control Engineering, Sept. 1986, pp. 294-302.
(6) H. Everett, Survey of Collision Avoidance and Ranging Sensors for Mobile Robots, Robotics and Autonomous Systems, vol. 5, pp. 5-67, 1989.
(7) A. Pugh, Robot Sensors, Vol. 1, Vision, Springer Verlag, Berlin, 1986.
(8) T. Strand, Optical Three Dimensional Sensing for Machine Vision, Optical Engineering, Jan. 1985, vol. 24, No. 1, pp. 33-40.
(9) N. Corby, Range Image Sensing and Processing for Robotic Control, Int'l Conference on Robot Vision and Sensory Control, Amsterdam, Oct. 29-31, 1985.
(10) T. Boult, A Survey of Some Three Dimensional Vision Systems, SIGART Newsletter, April 1985, No. 92, pp. 28-38.
(11) Z. Houkes, Distance Measurement by Stereo Vision, Int'l Conference on Robot Vision and Sensory Control, Amsterdam, Oct. 29-31, 1985.
(12) L. Bieman, Three Dimensional Machine Vision, Photonics Spectra, May 1988, pp. 81-92.
(13) R. Jarvis, A Perspective on Range Finding Techniques for Computer Vision, IEEE Trans. on Pattern Analysis and Machine Intelligence, vol. PAMI-1, No. 2, March 1983, pp. 122-139.
(14) P. Rives, L. Marce, Use of Moving Vision Sensors to Control Robots in an Unknown Universe, Int'l Conference on Robot Vision and Sensory Control, Amsterdam, Oct. 29-31, 1985.
(15) G. Dunlap, H. Shufeldt, Dutton's Navigation and Piloting, Naval Institute Press, May 1969.
(16) M. Beckerman, E. Oblow, Treatment of Systematic Errors in the Processing of Wide-Angle Sonar Sensor Data for Robotic Navigation, IEEE Trans. Robotics and Automation, vol. 6, No. 2, April 1990, pp. 137-145.
(17) S. Kuroda, A. Jitsumori, T. Inari, Ultrasonic Imaging System for Robots Using an Electronic Scanning Method, Robotica, vol. 2, 1984, pp. 47-53.

(18) D. Svetkoff, P. Leonard, R. Sampson, Techniques for Real-Time 3D Feature Extraction Using Range Information, Proc. SPIE paper 521.302, 1984.(19) M. Hersman, et al., Coherent Laser Radar Application to 3D
Vision, VISION-87 Conference, SME, Paper MS87-385:3-1, 1987.
(20) K. Lau, et al., Robot End Point Sensing Using Laser Tracking system, Proc. NBS and Navy NAV-CIM Robot Standards Workshop, Detroit, June 1985, pp. 104-111.
(21) A. Rahn, Gamma Irradiation of Resolvers, Internal Note AG-Mol Notiz 36/90, Dec. 18 1990.
(22) E. Abel, et al., NEATER, A Telerobot for the Nuclear Industry, ANS Conference on Robotics, Albuquerque, Feb. 91.
(23) V. Massaut, M. Decreton, R. Oris, Gastafiore (Gamma Irradiations of Sensors, Transducers and Optical Optics for Remote Operations, Irradiation Test of Fibre Optics and Proximity sensors for use in remote Handling and Maintenance, 16th Symposium on Fusion technology, London, Sept. 3-7, 1990.

ADVANCED TELEROBOTIC SYSTEMS SINGLE-MASTER MULTI-SLAVE MANIPULATOR SYSTEM AND CELLULAR ROBOTIC SYSTEM

TOSHIO FUKUDA and KAZUHIRO KOSUGE
Department of Mechanical Engineering
Nagoya University
Furo-cho, Chikusa-ku
Nagoya 464-01, JAPAN

ABSTRACT. This article proposes new telerobotic systems; one is the single-master multi-slave manipulator system and the other is the cellular robotic system. A Single-Master Multi-Slave Manipulator consists of a master arm maneuvered by an operator and slave arms used for real tasks. The master arm is used to control the task-oriented variables instead of slave arms so that the load of the operator is reduced. The Cellular Robotic System (CEBOT) is a new type of robotic system. It consists of many cells and autonomously reconfigures its structure according to given tasks. The system is adaptable to a variety of tasks.

I. Single-Master Multi-Slave Manipulator System

1. Introduction

A master slave manipulator system, in general, consists of a master arm maneuvered by an operator and a slave arm used for real operations. Advanced applications sometimes require several slave arms manipulated in coordination in order to increase the load capacity, rigidity and dexterity of the system. In the conventional framework of a single-master single-slave manipulator, slave arms in coordination are to be manipulated by an operator using master arms, each of which is corresponding to one of the slave arms. The simultaneous manipulation of master arms is a difficult job, because the operator has to coordinate the motions of the arms and control not only positions of the arms but also forces/moments applied to the manipulated object.

We have proposed a single-master multi-slave manipulator system based on the concept of "task-oriented control" [1], [2]. In this system, an operator is required to manipulate a single master arm with six degrees of freedom whose motion is relater to the task they want to do. This article summarizes the concept of the single-master multi-slave manipulator system.

In the following sections, we introduce the idea of the task-oriented control of a single-master multi-slave manipulator system, and propose a control architecture of the system. An experimental single-master multi-slave manipulator system is then developed. This system has two joysticks as a master, each of which has three degrees of freedom,

M. C. Becquet (ed.), Teleoperation: Numerical Simulation and Experimental Validation, 195–208.

and two slave arms, each of which has six degrees of freedom. Several experiments illustrate the concept of the system.

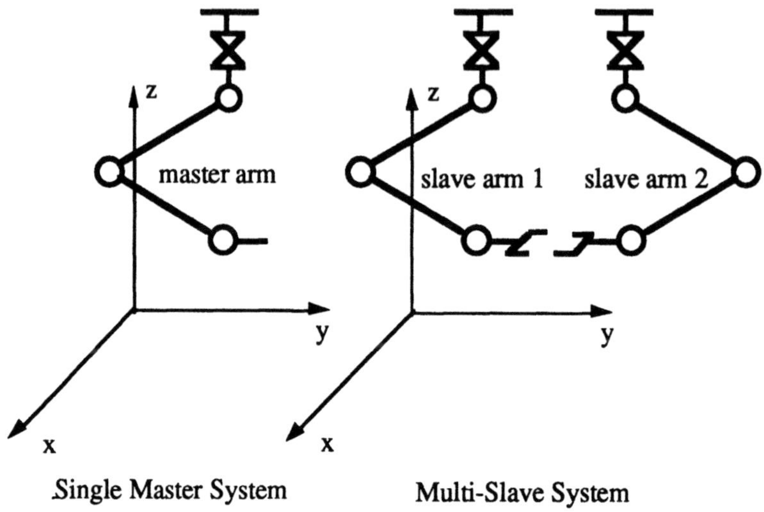

Fig.1 Example of Single-Master Multi-Slave System

2. Concept of Single-Master Multi-Slave Manipulator

A single-master multi-slave manipulator consists of a master arm manipulated by an operator and a set of slave arms used for a task. A master arm is a master device by which an operator can specify motions of six degrees of freedom. The coordination of slave arms increases the load capacity, the rigidity and the dexterity of the system as shown in fig.2. Let us consider how we can use a single-master to control slave arms in these examples. Two arms manipulating a heavy object that a single slave arm cannot handle in fig. 2 (a). What an operator of the system wants to do in this example is not to specify the motion of two arms but to specify the motion of the object with six degrees of freedom. Apparently, the motion of a single-master can specify the motion of the object. Two arms, in coordination, are using a tool having interactions with its environment in fig. 2 (b). In this example, what the operator wants to do is to specify the position/operation of the tool and the force/moment that the tool applies to its environment. A single-master can specify the position/orientation of the tool and the force/moment applied to the environment by employing a bilateral feedback. Two arms are assembling two pats, a bolt and a nut in fig. 2 (c). In this case, what the operator has to do is to control the difference in position and orientation of these two parts and the interface force/moment between these two parts. A single-master with six degrees of freedom can specify the difference in position and orientation as well as the interface force/moment by employing a bilateral feedback. Thus, in single-master multi-slave manipulator systems, we do not specify motions of slave arms using a master arm, but we use a master arm to specify task-oriented variables necessary for the execution of a task.

In some cases, such as fig.2(a) and fig.2(b), the information given by a master arm is not sufficient for the control of slave arms; take fig.2(a) and fig.2(b) for example, we have to specify the internal force/moment applied to the manipulated object which is not given by the motion of a single-master. Usually, the necessary information, which is not given by a master arm, can be determined by appropriate knowledge about the task.

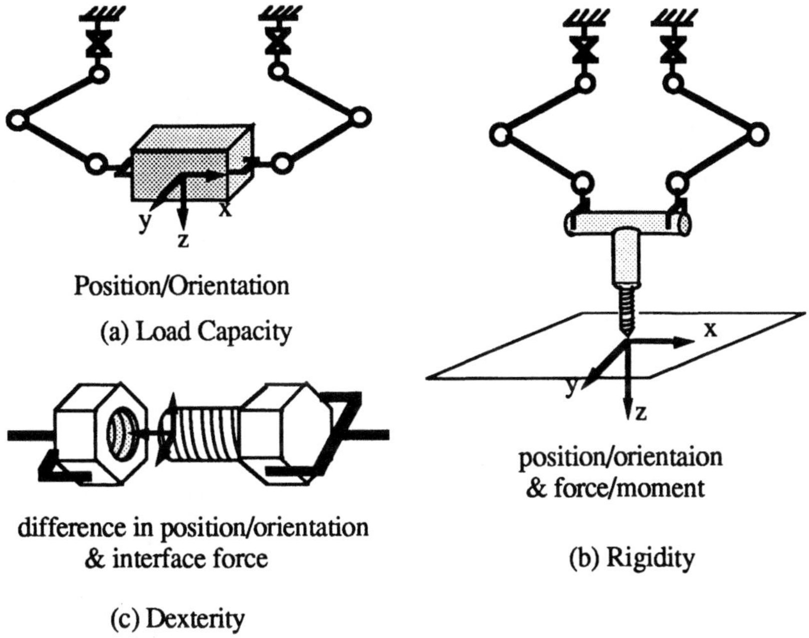

Fig.2 Tasks Using Multi-Manipulators

3. Concept of Task-Oriented Control

We review and extend the idea to the control problem of single-master multi-slave manipulators in this section. Motions of an object handled by manipulators to do a task, in general, consists of two types of motions, gross motions and fine motions. The gross motion is a preparatory motion for actual work, and the fine motion is a motion which comprises the work itself [6][7]. In the bilateral feedback of master-slave manipulators, the motion of a slave arm is controlled so as to follow the motion of a master arm and the motion of a master arm is controlled so as to feedback to the operator the interface force/moment between slave arm and its working environment. To do a task by means of such a force reflective master-slave manipulator, the operator has to specify both gross motions and fine motions. Other control algorithms have been developed to increase the realism with which an operator can generate fine necessary for tasks [9], etc. How to assist the operator to generate fine motions has been a key issue in such a framework of master-slave manipulator control.

The task-oriented control, as a new paradigm of master-slave manipulator control, generates task-oriented fine motions by itself instead of assisting an operator to generate them. Fine motions necessary for execution of a task are designed a priori, generated automatically based on the interface force/moment and superimposed on gross motions specified by an operator through a master arm. Other sensory information in addition to interface force/moment may be necessary and used for generating fine motions. In a single-master multi-slave manipulator system, task-oriented variables are controlled by a master arm. The task-oriented variables are concerned with the motion of an object, such as motion of a tool, the difference in position/orientation of parts, etc. Instead of specifying the task-oriented variables directly by a master arm, task-oriented control specifies fine motions necessary for task. How to design fine motions is an important issue for this control scheme. Much research has done concerned with the analysis and the design of fine motion [6][7][8]. Note that the fine motions are not necessary the mimic motions of a human operator. We can use artificial fine motions analyzed and designed for manipulators.

Some of the authors have proposed the control architecture of a master-slave manipulator system using a virtual internal model(VIM)[2]. A VIM is a reference model driven by sensory information [4], and was used to realize task-oriented functions which assist an operator to execute a task. It describes a dynamic relation between the motion of a master arm and that of a slave arm [2], for the conventional single-master single-slave manipulator system. We extend the idea to the task-oriented control of a single-master multi-slave manipulator system.

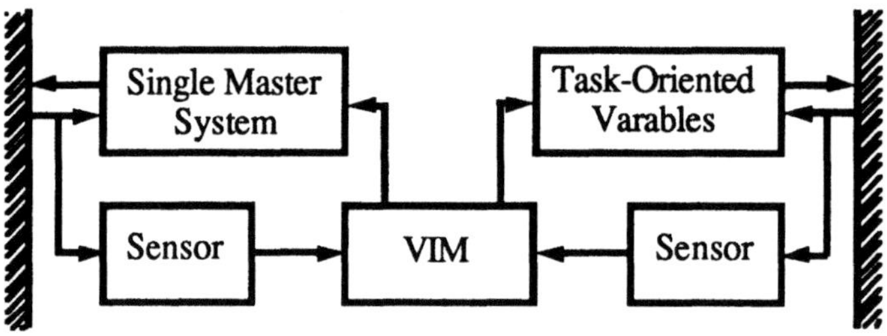

Fig. 3 Architecture of Task-Oriented Control

In a single-matter multi-slave manipulator system, the motion of a master arm specifies task-oriented variables. We use a VIM to describe the relation between the motion of a master arm and that of the task-oriented variables. The slave arms are controlled based on the task-oriented variables generated by a VIM. Many control algorithms have been proposed for the control problem of coordinating robot arms [10]-[17]. Here we apply a method proposed in [10] and [5] by authors. Refer to [10] for more detail of the control algorithm of slave arms. Fig.3 shows the control architecture of the system.

4. Experimental System and Conclusions

To illustrate the concept of the single-master multi-slave manipulator system, we have developed an experimental system with two slave arms. The system consists of two three-degree-of-freedom joysticks as a master arm and two six-degree-of-freedom robot arms (Nachi-Fujikosi) as slave system, one of the joysticks controls the position of the object or slave arms and the other one controls the orientation of them. A force sensor (Hitachi-Kenki) is attached to the end-effector of each slave arm. Through the experiments, the absolute coordinate is fixed to the base coordinate of the main slave arm as shown in fig.4. The calculation of the control laws of the two slave arms was executed within 1[msec] by high-speed micro-processor system(SONY) with Am 29332 for ALU and two Am 2935's for FPU, that is, the sampling rate of the slave arm control and force sensor was 1[kHz], on the other hand, the sampling rate of a reference signal through the master arm was 200[Hz].

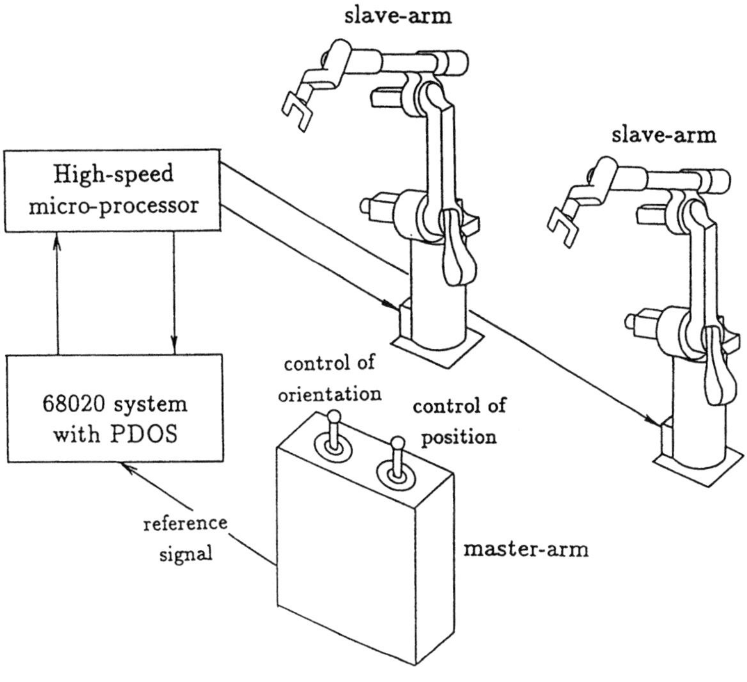

Fig.4 Experimental System

Fig.5 shows a palletizing task using this system. In this case, the task-oriented variables are the position/orientation of the manipulated box. The operator specified the gross motions of the box and fine motions we superimposed on them by the task-oriented control.

Some tasks require several slave arms operated in coordination. In the framework of the conventional master-slave manipulator systems, slave arms are to be controlled by master arms each of which corresponds to a slave arm. In this paper, we have proposed a single-master multi-slave manipulator system and its task-oriented control. In the proposed system, an operator controls task-oriented variables using a master arm. Slave arms are controlled on the task-oriented variables generated by a VIM according to the motion of a master arm and sensory signals such as interface force/moment between a manipulated object and its environment. With the proposed system, the operator no longer has to consider how to control the slave arms in coordination and can concentrate on their task during the operation.

We have developed an experimental single-master multi-slave manipulator system with two slave arms, each of which has six degrees of freedom, and illustrated the concept of our system. With the task-oriented control, we could execute a palletizing task, although the experimental system does not have a bilateral function and a master arm can specify the task-oriented variable up to six degrees of freedom. The use of the bilateral feedback as well as task-oriented functions will enable the system to deal with more complex tasks.

Fig.5 Box-Palletizing Task

II Cellular Robotic System

1. Introduction

Many robotic systems have been proposed and developed so far. Most of them have been developed for specific applications. It has been difficult to adapt the robotic systems to a variety of tasks and environments, since they cannot change their configuration by itself. A robotic system, which is dynamically reconfigurable, has been long desired. Fukuda and his group have proposed and studied the cellular robotic system(CEBOT) as a new type of intelligent robotic system(Fukuda, T.,IECON87). This article briefly reviews the concept of CEBOT as a next generation robotic system[18]-[28].

CEBOT has a similar structure to that of living creatures. CEBOT consists of cells or basic units of the system, and can build up modules. Each cell has its own simple function and intelligence with a data base, in the same way as the tissues and organs of a living creature, which are made up of cells. When CEBOT executes tasks, it is necessary to carry out task planning, optimal structuring and task execution automatically. Thus CEBOT can be considered an autonomous decentralized cooperative system with distributed intelligence. CEBOT has many advantages. It is a flexible system in terms of hardware and software; it is also fault tolerant, self-evolutionary, and self-organizing.

2. Concept of CEBOT (Cellular Robotic System)

CEBOT is an approach similar to biologic systems as shown in fig.6. The idea is to construct a system with a small number of kinds of simple components, which are able to connect and separate freely. In this way, a flexible structure can be realized. Many cells can connect to each other and make up modules. Many modules can form very complex systems. CEBOT has unique features similar to those of biologic organisms. Self-redundancy on the cellular level can be achieved by replacing defective components. If repair or replacement is not possible, other system modules or components can substitute for the function of defective parts. This flexible hardware structure of CEBOT, as well as the software system, is freely reconfigurable.

A single component of CEBOT, referred to as a cell, has its own intelligence, self-contained software, independent data base, sensor system, communication capabilities, a coupling mechanism to other cells, and a single mechanical function. Cells are classified in three groups by their mechanical function:

Group 1: Active actuating cells, including bending joint cells, rotating joint cells, sliding joint cells, end-effector cells, and mobile cells(wheel, crawler, and so on)
Group 2: Branching cells, orientation-changing cells, length adjustable arm cells between joints, power cells for heavy duty works, and others
Group 3: Work cells, end-effector cells, and special-purpose cells

Cell connections and separations are carried out automatically by mobile cells approaching other cells. Also joint and end-effector cells, which are already combined as a manipulator, can attach and detach other cells. Like biologic organisms, CEBOT is also organized in a hierarchicallystructureed system(fig. 7). On the lowest level (level 5) we fined cells with basic functions of groups 2 and 3. These cells have their own independent central processing unit (CPU), self-containing software, data base, and sensor system. They are principally completely autonomous intelligent units. However, the basic cells of level 5 are not able to move activity, which distinguishes them from mobile or joint cells of group1. These active cells are placed in level 4. Cells of level 4 and 5 can freely communicate with each other. Cells in level 4 are module masters. They form modules together with cells of level 5. When building up modules or large structures, the module master searches for its partner cells, automatically approaches, and docks to them.

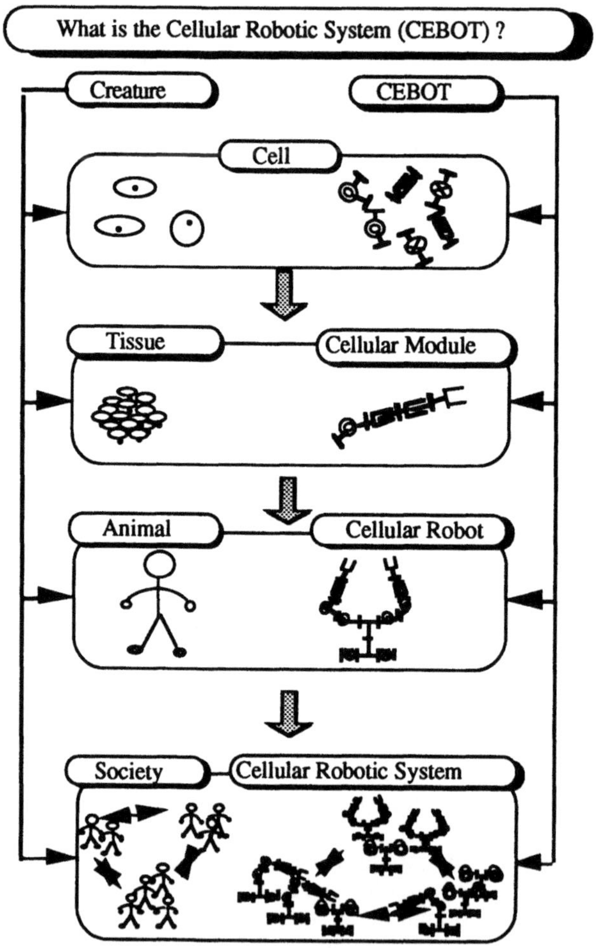

Fig. 6 Concept of CEBOT

Several modules can combine to form a large structure, which is controlled in the coordination level (level 3). In many applications, the coordination of uncoupled modules or mobile cells is desired, which is also controlled in level 3.

The structure-planning level (level 2) generates structures depending on given tasks. The required number of cells, their types, and their combination sequences are determined at this level.

Input commands to the structure-planning level come from the task-planning level (level 1), which is the highest level, corresponding to the central nervous system of biologic organisms.

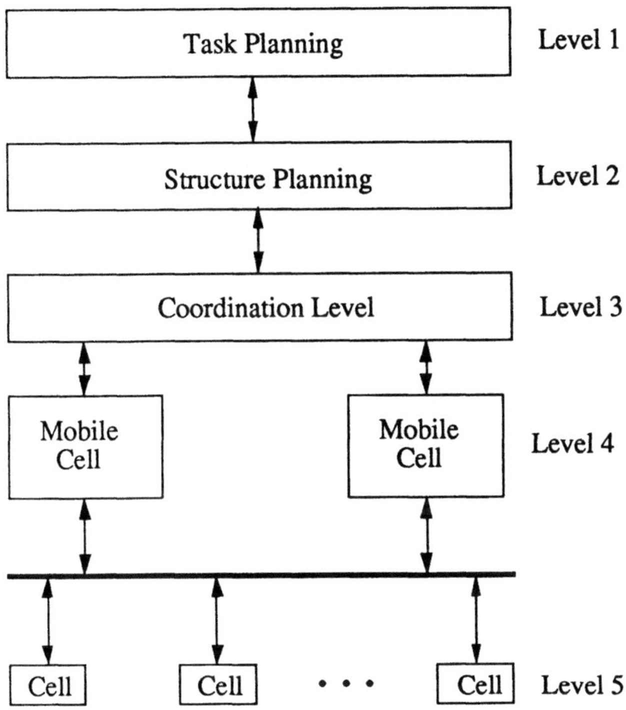

Fig.7 Hierarchical System Structure of CEBOT

3. Applications and Experimental Systems

In order to realize CEBOT, we have to solve problems such as automatic approach and docking/separation; obstacle avoidance; communication; optimal structuring; task planning; knowledge allocation etc. CEBOT can be used to form a manipulator system as a modular and also a mobile robotic system, depending on required tasks. We have developed three types of experimental models. Series II shown in fig. 8 has been developed to study the problem related to the hardware and communication. Series III shown in fig.9 has been developed to study the problem related to applications in space. The application fields of CEBOT are operational works in a closed world, space robotics with limited available resources, and others. Figs.10, 11 and 12 show its potential applications.

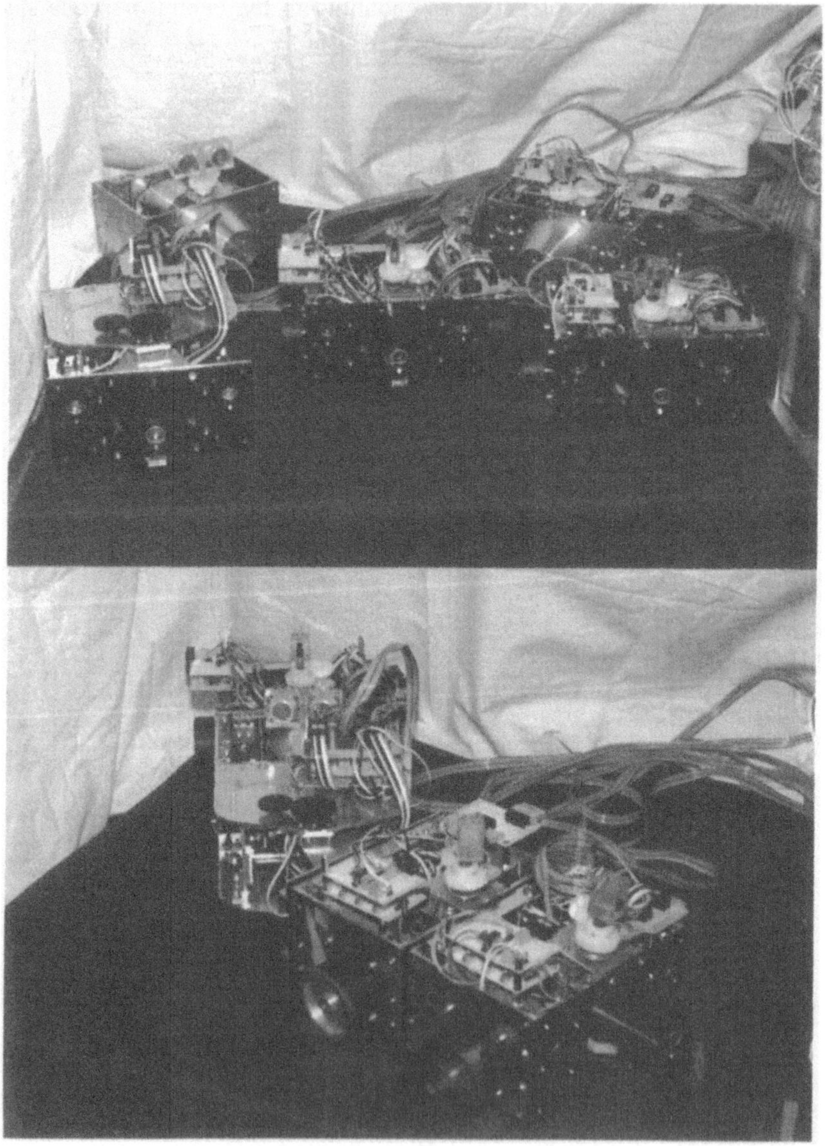

Fig.8 Basic Experiments Using Mark II

Fig.9 Mark III

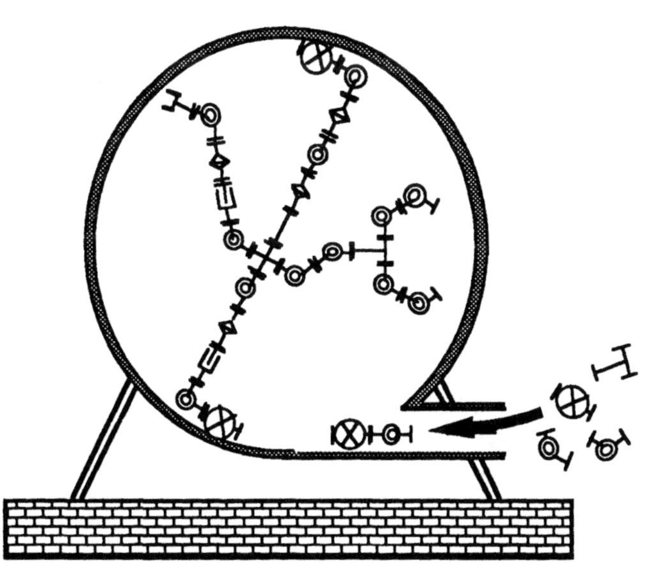

Fig.10 Application to Maintenance Robot

Fig.11 Application to Space Robot

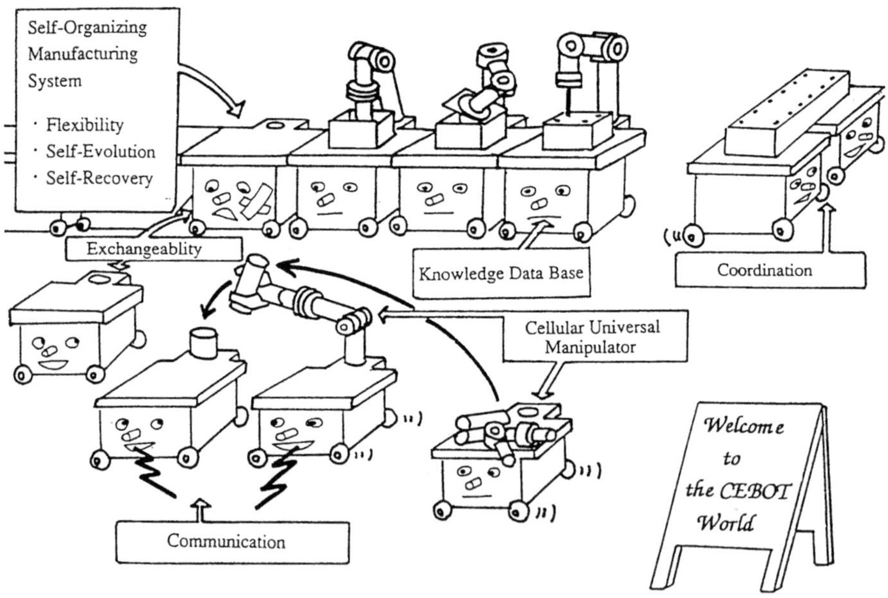

Fig.12 Application to Manufacturing System

References

[1] K.Kosuge, J.Ishikawa, K.Furuta and M.Sakai, "Control of Single-Master Multi-Slave Manipulator System Using VIM", Proc. of IEEE International Conference on Robotics and Automation, pp.1172-1177, 1990

[2] K.Furuta, K.Kosuge, Y.Shiote and H.Hatano,"Master-Slave Manipulator Based on Virtual Internal Model Following Control Concept", Proc. of 1987 IEEE International Conference on Robotics and Automation.pp.567-572,1987

[3] K.Kosuge, A.Sato and K.Furuta,"Task-Oriented Control of Master-Slave Manipulator", Proc.of 1990 Japan-USA Symposium on Flexible Automation, Kyoto, 1990

[4] K.Kosuge, K.Furuta and T.Yokoyama, "Virtual Internal Model Following Control of Robot Arms", Proc. of 1987 IEEE International Conference on Robotics and Automation, pp.1549-1554,1987

[5] K.Kosuge, J.Ishikawa, K.Furuta and M.Sakai, "Control of Single-Master Multi-Slave Manipulator System Using VIM", Proc. of 1990 IEEE International Conference on Robotics and Automation,1990

[6] D.E.Whitney, "A Survey of Manipulation and Assembly", Chapter 8, Robotics Science, edited by Michel Brady, MIT Press, Cambridge,1989

[7] D.E.Whitney, "Quasi-Static Assembly of Compliantly Supported Rigid Parts", Trans. of ASME, Journal of Dynamic System, Measurement and Control, Vol. 104, No.1, pp.65-77,1982

[8] H.Asada and Y.Kakumoto, "The Dynamic RCC Hand for High-Speed Assembly", Proc. of 1988 IEEE International Conference on Robotics and Automation, pp.120-125,1988

[9] Blake Hannaford, "A Design Framework for Telepoerators with Kinesthetic Feedback",IEEE Trans. on Robotics and Automation, Vol.RA-5, No.4, pp.426-434,1989

[10] K.Kosuge, M.Koga, K.Furuta,K.Nosaki, "Coordinated Motion Control of Robot Arm based on Virtual Internal Model", Proc. of 1989 IEEE International Conference on Robotics and Automation, pp.1097-1102,1899

[11] E.Nakano, S.Ozaki,T.Ishida and I.Kato, "Cooperation Control of the Anthropomomorphous Manipulator 'MELARM'", Proc. of the 4th International Symposium on Industrial Robots, pp.251-260, Tokyo 1974

[12] S.Kurono, "Coordinated Computer Control of a Pair of Artificial Arms", Biomechanism 3, The Tokyo University Press, pp.182-193,1975

[13] Y.Nakamura, K.Nagai and T.Yoshikawa, "Mechanics of Coordinative Manipulation by Multiple Robotic Mechanisms", Proc. of 1987 IEEE Conference on Robotics and Automation, pp.991-998,1987

[14] M.Uchiyama and P.Dauchez, "A Symmetric Hybrid Position/Force Control Scheme for the Coordination of Two Robots", Proc. of 1988 IEEE International Conference on Robotics and Automation, pp.350-356, 1988

[15] S.Hayati, "Hybrid Position/Force Control of Multi-Arm Cooperating Robots", Proc. of 1986 IEEE International Conference on Robotics and Automation, pp.82-89,1986

[16] T.J.Tarn, A.K.Bejczy, and X.Yun, "Design of Dynamic Control of Two Cooperating Robot Arms; Closed Chain Formulation", Proc. 1987 IEEE International Conference on Robotics and Automation, pp.7-13,1987

[17] Y.F.Zheng and J.YS.Luh, "Control of Two Coordinated Robots in Motion", Proc. 24th IEEE Conference on Decision and Control, Vol.1, pp.1761-1765,1985

[18] Fukuda, T. and Nakagawa, S.:"A Self Recognized Robotic System with Cell Structure", Trans. of IEE Japan, Vol.107-C, No.11, pp.1019-1026, 1987

[19] Fukuda, T. and Nakagawa, S. "A Dynamically Reconfigurable Robotic System (Concept of a System and Optimal Configuration)", Proc. IECON'87, 588-595, 1987
[20] Fukuda, T. and Nakagawa, S. "Approach to the Dynamically Reconfigurable Robotic System", Journal of Intelligent and Robotic Systems 1, (1988), 55-72
[21] Fukuda, T., Nakagawa, S., Kawauchi, Y and Buss, M.:"Self Organizing Robots Based on Cell Structures - CEBOT", Proc. 1988 IEEE International Workshop on Intelligent Robots and Systems (IROS'88),pp.145-150
[22] Fukuda, T., Nakagawa, S., Kawauchi, Y and Buss, M.:"Structure Decision Method for Self Organizing Robots based on Cell Structures-CEBOT", Proc. 1989 IEEE International Conference on Robotics and Automation Vol. 2, pp. 695-700
[23] Fukuda, T. Kawauchi, Y and Buss, M. "Communication Method of Cellular Robotics CEBOT as a Selforganizing Robotic System," Proc. 1989 IEEE International Workshop on Intelligent Robots and Systems (IROS'89), pp. 291-296, (1989)
[24] Fukuda, T and Kawauchi, Y. "Cellular Robotic System(CEBOT) as One of the Realization of Self-organizing Intelligent Universal Manipulator," Proc. 1990 IEEE International Conference on Robotics and Automation pp. 662-667, (1990)
[25] Fukuda, T. Kawauchi, Y. and Asama, H. "Analysis and Evaluation of Cellular Robotics(CEBOT) as a Distributed Intelligent System by Communication Information Amount," Proc. 1990 IEEE International Workshop on Intelligent Robots and System (IROS'90), pp. 827-834, (1990)
[26] Fukuda, T. Ueyama, T and Itoh, Y. "Task Planning System for Determining the Form of Cellular Robotics(CEBOT)," Proc. of Pacific Rim International Conference on Artificial Intelligence'90, pp. 841 - 846, (1990)
[27] Fukuda, T. Ueyama, T. Arai, F. "Control Strategy for a Network of Cellular Robots - Determination of a Master Cell for Cellular Robotic Network based on a Potential Energy -," Proc. 1991 IEEE International Conference on Robotics and Automation , 1991
[28] Fukuda, T. Ueyama, T. Arai, F. "Communication Strategy for Network of Cellular Robotic System (CEBOT) - Mastering method for single and two tops structures -," Proc. 1991 IEEE International Conference on Advanced Robotics (ICAR'91), pp.528-532, (1991)

INTRODUCTION TO

ROBOTICS AND COMPUTER VISION

J.E. BESANÇON
Scientific Expert, TELEMAN Programme
Commission of the European Communities, DG XII
Rue de la Loi, 200
1049 Bruxelles
Belgique

ABSTRACT. This introduction is intended to provide a fundamental knowledge of Computer Vision Science; the first part describes some useful and basic notions of mathematical transforms (Fourier), image sampling, digitizing and filtering techniques. Then, definitions of image segmentation and restoration, figure extraction and object classification are given as well as notions of optical flux, artificial neural networks and workspace representation. The second part deals with practical applications such as robot navigation and object recognition in a scene.

I. Introduction

Image processing was really born in the USA in the sixties when NASA programmes concerning the discovery of the earth's environment induced a fast development of all vision by cameras embedded in satellites spinning around earth and moon and transmitting by radiowaves digital images to earth stations (RANGER lunar probe, SURVEYOR mission and finally the APOLLO flight in July 1969). Image processing then became a technology which incorporated many other existing and proven techniques and its applications have encompassed different fields such as geophysics, astrophysics, robotics (mobile robot vision), computer graphics, computer aided design, medical image data analysis, character recognition, military applications (battlefield vision, target acquisition, distance evaluation, etc), manufacturing processes automation, product inspection; and that was how all the new technologies which have implemented digital imaging systems created a new science: Computer Vision.

II. Basic and theoretical notions of Computer Vision

A. MATHEMATICAL TRANSFORMS

1. Fourier transform

With a space variable x and a frequency variable u the Fourier transform F(u) of a function f(x) is given by:

M. C. Becquet (ed.), Teleoperation: Numerical Simulation and Experimental Validation, 209–241.

$$\mathcal{F}\{f(x)\} = F(u) = \int_{-\infty}^{+\infty} f(x) \exp(-2\pi iux)\, dx$$

conversely:

$$\mathcal{F}^{1}\{ F(u) \} = f(x) = \int_{-\infty}^{+\infty} F(u) \exp(2\pi iux)\, dx$$

If f(x) is a real function then F(u) is a complex function represented by the sum of real and complex terms R and I :

$$F(u) = R(u) + iI(u)$$

The signal spectrum is : $|F(u)| = (R^2(u) + I^2(u))^{1/2}$ and

the energy spectrum : $E(u) = [F(u)]^2$

<u>example</u>

$$F(u) = \int_0^X A \exp(-2\pi iu)\, dx \qquad \text{with } f(x) = A$$

$$F(u) = \frac{-A}{2\pi iu} | \exp(-2\pi iux)|_0^X$$

and $|F(u)|$ can be written :

$$| F(u) | = AX \left| \frac{\sin \pi uX}{\pi uX} \right| = AX\, sinc\, (\pi uX)$$

where sinc is the cardinal sine function.

The Fourier transform of f(x,y), where x and y are the space variables in the plan and u and v the frequential variables, can be written as

$$F(u,v) = \iint_{-\infty}^{+\infty} f(x,y) \exp\, [-2\pi i(ux+vy)]\, dxdy$$

$$| F(u,v) | = [R^2(u,v) + I^2(u,v)]^{\frac{1}{2}} \qquad \text{is the spectrum of } f(x,y)$$

$$E(u,v) = R^2 + I^2 \quad \text{is the energy spectrum.}$$

In the current applications the Fast Fourier Transform, FFT, is a popular and fast algorithm especially used in computer vision.
The general formulation of mathematical transforms T(u) is expressed as follows :

$$T(u) = \sum_{x=0}^{N-1} f(x)\, g(x,u)$$

the term g(x,u) being the **kernel** of f(x).

Conversely the function f(x) can be extracted from T(u)

$$f(x) = \sum_{u=0}^{N-1} T(u)\, h(x,u)$$

the term h(x,u) being the inverse kernel.
In the two-dimensional space

$$T(u,v) = \sum_{x=0}^{N-1} \sum_{y=0}^{N-1} f(x,y)\, g(x,y,u,v)$$

When the kernel is symmetric and separable, it can be written as :

$$g(x,y,u,v) = g_1\,(u,x)\; g_2\,(v,y).$$

In the Fourier transform the kernel is typically separable

$$g\,(x,y,u,v) = \exp\,[-2\pi i\,(ux + vy)].$$

In the field of vision the most current and geometrically invariant transforms used, with the exception of FFT, are : HADAMARD, WALSH, HOTELLING or KARHUNEN-LOEVE, Discrete Cosine and MELLIN (for pattern recognition applications).

Example : Figure A-1 shows two different digital images, i.e. 2 two-dimensional functions and their Fourier spectrum. We note that a Fourier transform is not a legible image; the advantage of Fourier transforms is the mapping of the real plan into the frequency domain; all the problems concerning image convolution and correlation are easily achieved in the FFT domain; the most important applications are pattern recognition with the help of correlation and image restoration by filtering techniques which are equivalent to convolution.

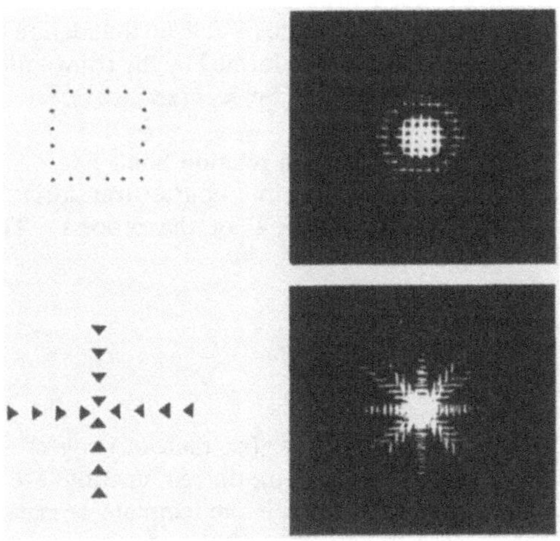

Fig. A-1 : two-dimensional digital images and their Fourier spectra.

2. Fourier transform applications

a. Convolution product; this is given by :

$$f(x) \otimes g(x) = \int_{-\infty}^{+\infty} f(\alpha)\, g(x-\alpha)\, d\alpha$$

where α is a dummy variable of integration.
If F(u) is the Fourier transform of f(x) and G(u) of g(x), then f(x) $\otimes$ g(x) is equivalent to F(u) G(u); conversely, f(x) g(x) is equivalent to F(u) $\otimes$ G(u).

The current application of the convolution is the sampling of a truncated signal only defined inside the interval $|-w, w|$, fig A-2; the convolution provides tools for analyzing the choice of the sampling **pitch**.

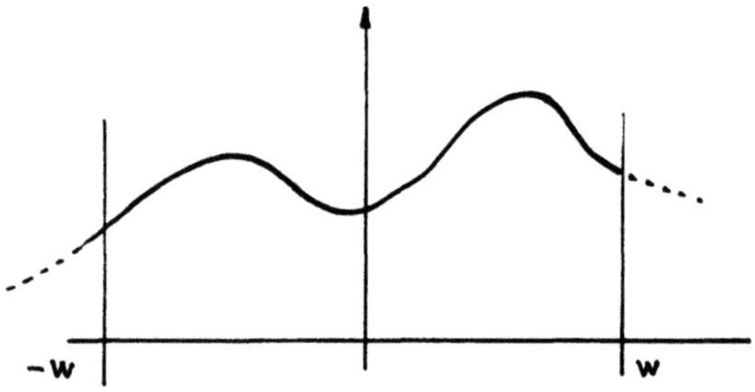

Fig. A-2

Another important application is the frequency filtering of a digital image; edges and sharp transitions in the gray levels of an image contribute to the high-frequency content in its Fourier transform; therefore blurring for example, can be achieved via the frequency domain by attenuating a specified range of frequency components in the image FFT. The foundation of frequency domain techniques is the convolution. Let g(x,y) be an image formed by the convolution of an image f(x,y) and a "position-invariant operator" h(x,y), that is: g(x,y) = h(x,y) f(x,y).

Then, we obtain that the following frequency-domain relation holds :
G(u,v) = H(u,v)F(u,v) , where G, H and F are the Fourier transforms of g, h and f.
The transform H(u,v) is called the "transfer function" of the process. This problem will be furthermore analysed in the following chapters.

b. Correlation product; this is given by :

$$f(x) \circ g(x) = \int_{-\infty}^{+\infty} f(\alpha)\, g(x+\alpha)\, d\alpha$$

If F(u) is the Fourier transform of f(x) and G(u) of g(x), then f(x) $\circ$ g(x) is equivalent to F(u) G^*(u); conversely, f(x) g^*(x) is equivalent to F(u) $\circ$ G(u); starred functions are complex conjugates. One of the most usual applications of the correlation is the template or mask matching. Given a digital image f(x,y), suppose that we wish to determine if it contains a region which is similar to some region w(x,y); the solution to this problem is to perform a correlation between w(x,y) and f(x,y).
The correlation product between these two functions is given by :

$$R(m,n) = \sum_{x} \sum_{y} f(x,y)\, w(x-m, y-n)$$

where m = 0, 1, 2..., M-1 and n = 0, 1, 2..., N-1; M and N being the size of the image f(x,y). The summation is taken over the image region where w is defined.
The procedure is illustrated in Fig.A-3; for any value of (m,n) inside f(x,y) we apply the equation given above, in order to obtain one value of R. As m and n are varied, w(x,y) moves around the image area and we obtain the function R(m,n).
The maximum value of R(m,n) then indicates the position where w(x,y) best matched f(x,y).

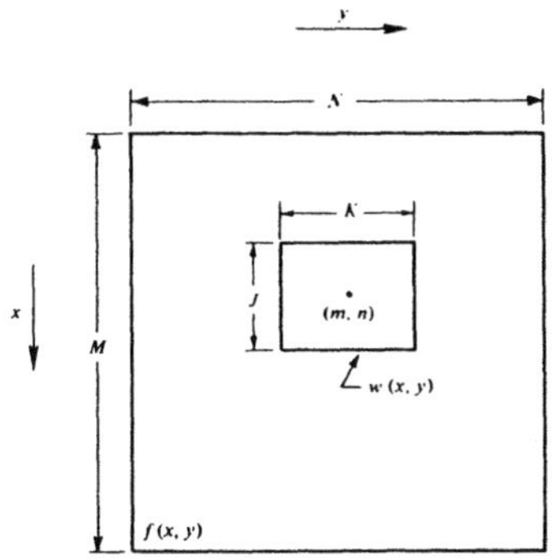

Fig. A-3 : the size of the mask is J x K; arrangement for obtaining the correlation of f(x,y) and w(x,y) at a given point (m,n).

It is usual to introduce a normalizing factor into R(m,n) by dividing the expression by :

$$[\sum_{x} \sum_{y} f^2(x,y)]^{1/2}$$

It is noted that accuracy will be lost for values of m and n near the edges of f(x,y), with the amount of error being proportional to the size of w(x,y).
The correlation and convolution products are carried out in the frequency domain via FFT algorithms.

B. IMAGE SAMPLING AND DIGITIZING

1. Image model

As used in this lecture, the term image refers to a two-dimensional light intensity function denoted f(x,y), where the value f at spatial coordinates (x,y) gives the brightness of the image at the point. In order to take a suitable form for computer processing, an image function f(x,y) is digitized spatially and in amplitude. Digitizing operation of spatial coordinates is referred to as image sampling while amplitude digitizing is called gray-level quantization. These two operations are carried out by scanning the original image line by line. (Scanning is adapted to the support of the image : spot on paper, electron-beam in a cathode ray tube, shift-register in a CCD camera matrix.) The beam is defined by geometrical dimensions or **aperture**.
The analog and continuous image f_c = f(x,y) is approximated by samples $\delta(x,y)$, or Dirac signals, and is written in an array of N x N discrete quantities as :

$$f(x,y) = \begin{vmatrix} f(0,0) & f(0,1) & \ldots\ldots\ldots\ldots f(0,N-1) \\ f(1,0) & f(1,1) & \ldots\ldots\ldots\ldots f(1,N-1) \\ . & & \\ . & & \\ f(N-1,0) & & \ldots\ldots\ldots\ldots f(N-1,N-1) \end{vmatrix}$$

Each element of the array is referred to as a picture element or **pixel**. In the above digitizing process it is common practice to let the quantities f be integer powers of two. If G denotes the number of gray-level and N the number of pixel by line, it is usual to write:

$$N = 2^n \quad \text{and} \quad G = 2^m$$

Then, the number of bits required to store a digital image is given by : $b = N^2 \times m$, i.e. N^2 pixels of m bits. For example, a 512 x 512 image with 64 gray levels (m = 6) requires 1 572 864 bits or 256 Kbytes.

The sampling operation with respect of a regular pitch P, consists in recording the amplitudes of f_s only in the multiple points of P. There is one solution to face when dealing with samples sequences, which results from sampling i data items at equal intervals of time and space. The Shannon sampling theorem sets the minimum rate for sampling a bandwidth-limited continuous signal so it can be uniquely recovered from its samples. The theorem states that the sampling frequency f_n must be at least twice the bandwidth of the signal f_s, i.e. the largest frequency in the signal. The sampling frequency is called the Nyquist frequency and $f_n = 2f_s$. If the sampling theorem is not obeyed, **aliasing** results; this occurs when a continuous signal is sampled at such a rate that the translated versions of its spectrum created by the sampling process overlap in the frequency domain. Since the image is approximated by a set of discrete intensity values f(x,y) stored in a squared array, the degree of discernable detail or **resolution** is a question relevant to the quality of digitized images; in other aspects the question is to ask how many samples and gray levels are required for a good approximation; for example, the future HDTV system aims at 1024 x 1024 images with colour levels (3 times 64 levels) displayed on an 80 cm. diagonal length screen.

2. Image sensors

Image sensors or electronic cameras perform mapping of the three-dimensional space (emitter) on to the two-dimensional image plan (receiver); they convert light into an electric signal; they provide both B & W and colour images and mostly fall into two different categories of components :

- Video tubes : vacuum tubes using electron beam.
- Solid State Video cameras using advanced silicon technology.

The first family representing tubes is defined by the photocathode interlace scanning with the help of an electron beam coming out of a heated gun; the simplified and portable model of the first Orthicon tube series is now the Vidicon. Plumbicon is a Vidicon equipped with lead oxide photocathode; its sensitivity is higher and remanence lower .

The second family concerns all the Solid State cameras and essentially employs interline transfer charge-coupled devices (CCD) which contain an optical sensor or cell and both vertical and horizontal registers; transfer gates are arranged on the left and right sides of the register and also along the cells. The incident light is converted into the signal charge by the cells and then stored; stored signals are read out to the register when the gates are opened. There also exists a second variety of solid state sensor which uses charge-injected devices (CID); charges are stored in micro-capacitors; signal read-out is accomplished by addressing the capacitors or an X-Y basis. The Solid State cameras have a photosensor array form or are one-line devices for streak-camera use.

Some definitions concerning all types of cameras are given below :

- definition : feature of pixel number by surface unit; for example in a CCD one-line camera, cells are 7 μm wide (2 μm are dedicated to insulation and channel stoppers);
- blooming : process of charge overspilling from one cell to the next; this creates temporary camera blinding when the light is too high.
- dark current : residual signal when camera cells or photocathodes are not lighted.
- DSNU : Dark Signal Non Uniformity.

A comparison between different types of photocathode tube and CCD array colour-cameras is given in the following table :

	CCD	VIDICON-like TUBE
Lifetime; reliability	long; stable	damage due to election beam heating
Weight and dimensions	reduced in weight and size	long size; heavy weight due to deflection coils.
Remanence	virtually nil	not avoidable; due to photo-conductor layer
Blooming	highly resistant	weakness of tube
Gamma	≈ 1	weaker (non linearity)
Distortion	nil	high along edges
Pre-heating	nil	20 seconds delay
Consumption	low	significant due to electron gun heating
Use in E-M field	without side effect	not negligible
Sensitivity	up to 1-5 lux for F:1-4	less
Resolution	excellent up to 450 lines excellent with one-line devices	excellent up to 650 lines
Colour	lack of contrast	perfect quality
Applications	industry, robotics	Television networks
Pixel size, nb.	15 x 12 μm; 440 000	525 lines
Scanning frequency	14.19 MHz	15.75 kHz/line

C. OPTICAL FLOW

1. Definition

Object shape and distance information can be perceived by humans when they are viewing moving random dot fields in the absence of non motion cues with only one eye. This applies to movement of individual objects in a random dot scene and to movements of the head when viewing a dot scene under conditions that simulate visual movement parallax. In order to understand and utilize such motion information, the first idea is to process time varying images to obtain an intrinsic optical flow function. The optical flow is a vector flow field that records, pointwise, the instantaneous velocity of gray-scale pattern displacements in the plane of image formation.

2. Real and apparent optical flows

Consider a set of curves (c_t) where t is time; these curves are supposed to be projections, through an optical system such as a camera, on to the planar space R (image-plane x,y) of a curve (C) moving in the 3-dimensional space. In the frame (x,y,t) the set of curves sweeps over a surface (Σ) which is defined as the set of points (c_t, t).

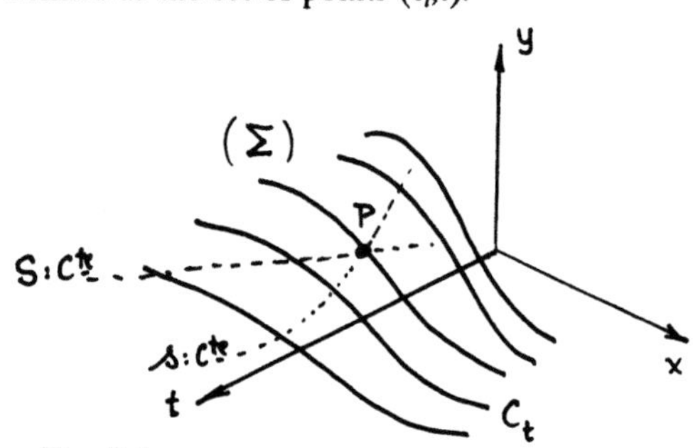

Fig. C-1

The observed curve (c_t, t), at a given time t, has a calculable arc length s; therefore (c_t) has 2 parameters, s and t, and is formed by the set of points $m_t(s)$. The corresponding points P on (Σ) are given by the vector $\vec{P}$.

$$\vec{P} = \begin{Bmatrix} m_t(s) \\ t \end{Bmatrix}$$

Key observations :

a) the arc length s = s (S,t) of (c_t) is a function of arc length S taken on the curve (C).

b) the 2 parameters S and t are used to define the surface (Σ) in the neighbourhood of P; in general the function s(S,t) is unknown.

Now, consider on the surface (Σ) all the curves defined by the following conditions : s = constant and S = constant, i.e., a point S_o is fixed on the arc S and s_o on the arc s. These curves are very often different and their projections on the plane R in the direction of t axis are useful for the following definition of the flows $\vec{V}_r$ and $\vec{V}_a$.

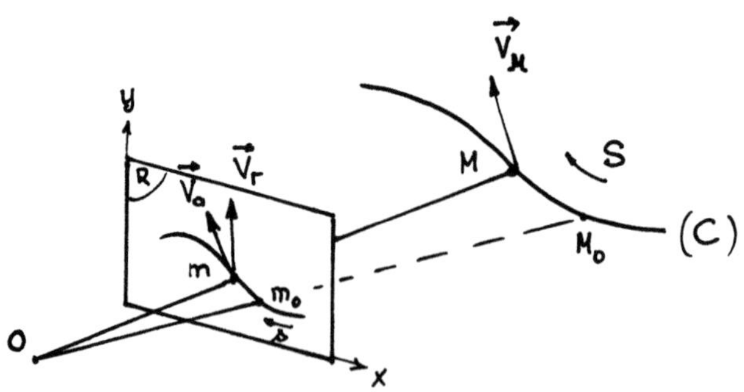

Fig. C-2

M_o is a point chosen on the curve (C) for S = S_o; when (C) is moving, M_o follows a trajectory (C_{Mo}) in the roomspace and its image m_o, corresponding to s_o, follows a trajectory (c^r_{mo}) on the plane R; the curve (c^r_{mo}) is the projection on the plane R, in the direction of t axis, of the curve defined by S_o on (Σ) : this is the real trajectory of m_o.
Now consider the same projection of another point m_o defined by s_o on (Σ); the corresponding curve on R is (c^a_{mo}); this is the trajectory of m_o taken in c_i; this curve is called the apparent trajectory of m_o. The velocities V_r and V_a, or real and apparent optical flows, are indicated in Fig. C-2.
V_r, real optical flow, is the partial derivative/t of m when the arc (S) keeps a constant value S_o.
V_a, apparent optical flow of point m is the velocity defined as the partial derivative/t when arc s keeps a constant value s_o; this is the velocity of m_0 in the image-plane R.
V_r and V_a are in general distinct.

3. Spatiotemporal gradient approach

This approach for computing optical flow uses first-order spatial and temporal gradients of a time-varying image and, at each image point, estimates the component of motion in the direction of maximally increasing gray scale intensity. (Ref.1).
Assume E (x,y,t) is an analytical function describing the gray-value pattern at the point (x,y) and at the time t.(gray value is equivalent to brightness at a given point); when the pattern moves the observer linked to the camera frame O (fig.C-2) verifies the brightness of real flow to be constant, so that :

$$\frac{dE}{dt} = 0$$

Using the chain rule for differentiation we get :

$$\frac{\delta E}{\delta x}\frac{dx}{dt} + \frac{\delta E}{\delta y}\frac{dy}{dt} + \frac{\delta E}{\delta t} = 0$$

Noting that the velocity components u and v are respectively dx/dt and dy/dt, then the last equation becomes :

$$u\frac{\delta E}{\delta x} + v\frac{\delta E}{\delta y} + \frac{\delta E}{\delta t} = 0$$

u and v are components of the apparent velocity in the image-plan R; in the following this velocity is simply called $\vec{V}$.
Introducing the additional abbreviations E_x, E_y and E_t for the partial derivatives of E with respect to x, y and t, we can rewrite the equation as :

$$u\,E_x + v\,E_y = -\,E_t$$

$$\text{or}\quad \vec{V}.\vec{\nabla}E = -E_t \qquad \text{with } \vec{\nabla} : \text{nabla, vector-symbol of gradient}$$

In this expression there are 2 unknown values u and v for only one equation, therefore, E_t is called the **first constraint** term. Thus, the component of the movement in the direction of the brightness gradient $\vec{\nabla}E$ equals :

$$-\frac{E_t}{(E_x^2 + E_y^2)^{1/2}} = |\vec{V}| = (E_x,\ E_y)$$

We cannot determine the component of the movement in the direction of the iso-brightness contours, at right angles to the brightness gradient, fig.C-3; as a consequence, the flow velocity (u,v) cannot be locally computed without introducing an additional constraint.

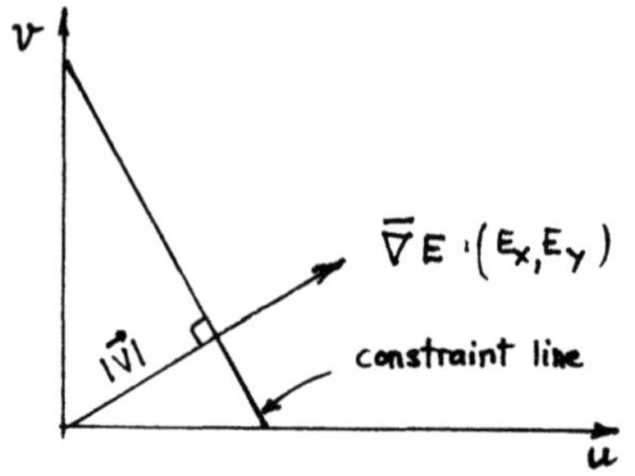

Fig. C-3

4. Solutions

One way to express the additional constraint is to minimize the square of the magnitude of the gradient of the optical flow velocity, (ref.1) :

$$\left(\frac{\delta u}{\delta x}\right)^2 + \left(\frac{\delta u}{\delta y}\right)^2 \quad and \quad \left(\frac{\delta v}{\delta x}\right)^2 + \left(\frac{\delta v}{\delta y}\right)^2$$

This approach is long and tedious; we propose another method based on an approximation concerning the brightness E. Several authors (ref.2) consider that instead of keeping E as a constant value, the image velocity field is stationary, so that :

$$\frac{d}{dt}\vec{\nabla}E = 0$$

With the help of the relation :

$$\frac{dE}{dt} = \left(\frac{\partial}{\partial t} + \vec{v}\cdot\vec{\nabla}\right) E = 0$$

the stationary condition becomes :

$$\left(\frac{\partial}{\partial t} + \vec{V}\cdot\vec{\nabla}\right) \vec{\nabla}E = 0$$

and let come into sight the gradient components as :

$$\frac{d\vec{\nabla}E}{dt} = \begin{vmatrix} \frac{d}{dt}E_x \\ \frac{d}{dt}E_y \end{vmatrix} = \begin{vmatrix} E_{xx}u + E_{xy}v + E_{xt} \\ E_{xy}u + E_{xx}v + E_{yt} \end{vmatrix} = 0$$

The system in a matrix form is given by

$$\begin{vmatrix} E_{xx} & O \\ O & E_{yy} \end{vmatrix} \begin{vmatrix} u \\ v \end{vmatrix} = \begin{vmatrix} -E_{xt} \\ -E_{yt} \end{vmatrix}$$

where, for example, E_{xx} is the second-order x derivative of E.

Transforming the matrix product in the following shape :

$$\boldsymbol{H} \cdot \vec{V} = -\frac{\partial}{\partial t} \vec{\nabla} E$$ with H : Hessian matrix

and introducing the determinant Δ of the multiplicand matrix of the transpose vector $| u,v |^T$, we get:

$$\begin{vmatrix} u \\ v \end{vmatrix} = -\frac{1}{\Delta} \begin{vmatrix} E_{yy} & 0 \\ 0 & E_{xx} \end{vmatrix} \begin{vmatrix} E_{xt} \\ E_{yt} \end{vmatrix}$$

and finally : $\vec{V} = -1/\Delta (\boldsymbol{H})^{-1} \frac{\partial}{\partial t}\, grad\, E$

The optical flow given by the last expression is the apparent velocity and does not strictly correspond to the value given by the first expression of V_a but it is a robust and well approximated calculation.

(ref. 1 : B.K.P. HORN & B.G.SCHUNK; "Determining Optical Flow"; Artificial Intelligence 17, 1981, pp. 185-203. - ref. 2 : A. VERRI, F. GIROSI & V. TORRE;"Constraints for the Computation of Optical Flow"; Proc. of IEEE Workshop on Visual Motion, Irvine CA, 1989, pp.116-124)

D. IMAGE PREPROCESSING AND SEGMENTATION

1. Image preprocessing

The approaches discussed in this chapter are divided into two categories : frequency-domain methods where the processing techniques are based on modifying the Fourier transform of an image like filtering and restoration; spatial-domain which refers to the real image plane, i.e. pixel array; in the last category techniques are based on direct manipulation of the pixels, like histogram equalization, sharpening by gradient, Hough transform and general methods of edge detection.

a. Filtering techniques

As said before, blurring can be achieved by attenuating high frequency (HF) components in the Fourier transform of a given image because edges and other sharp transitions such as noise contribute heavily to the high-frequency spectrum. The attenuation is implemented by filtering the HF spectrum and the remaining information pass through in the low-frequency (LF) range.
This method is commonly referred to as low pass filtering and is performed in the Fourier space as:

$$G(u,v) = H(u,v)\, F(u,v)$$

which represents in the real space the convolution product of the real image f(x,y) by h(x,y); here h(x,y) is the inverse Fourier transform of H. H(u,v) is referred to in this context as a filter transfer function. Several low pass filtering solutions are discussed there and in all cases, the filter-functions are know as zero-phase-shift-filters because they do not alter the phase of the transform.

a-1. Ideal Lowpass filter

Here is a 2D ideal low-pass filter

$$H(u,v) = \begin{Bmatrix} 1 \ if \ D(u,v) \leq D_0 \\ 0 \ if \ D(u,v) > D_0 \end{Bmatrix}$$

Where D_o is a specified positive quantity and D(u,v) the distance from point (u,v) to the origin of the frequency space, that is:

$$D(u,v) = (u^2 + v^2)^{1/2}$$

a-2. Butterworth filter

The transfer function of this lowpass filter of order n presents a cut-off frequency locus at a distance D_o from the origin and is defined by the relation :

$$H(u,v) = \frac{1}{1 + [D(u,v,/D_o]^{2n}}$$

a-3. Exponential filter

This is a smooth filter presenting a cut-off frequency locus at a distance D_o from the origin and is given by :

$$H(u,v) = exp - [D(u,v)/D_o]^n$$

a-4. Highpass filter

Since edges and other abrupt changes in gray-levels are associated with HF components, image sharpening can be achieved in the frequency domain by a highpass filter process which attenuates low frequency components without disturbing high frequency information.
The ideal highpass filter is one whose transfer function satisfies the relation:

$$H(u,v) = \begin{Bmatrix} 0 \ if \ D(u,v) \leq D_0 \\ 1 \ if \ D(u,v) > D_0 \end{Bmatrix}$$

where D_o is the cut-off distance measured from the origin of the frequency and D(u,v) is given as before.
There exists also Butterworth and exponential forms; in the last one given by

$$H(u,v) = exp - [D_o/D(u,v)]^n$$

when D(u,v) = D_o, we observe that H(u,v) = 1/e; a simple modification forces H(u,v) to be equal to $1/\sqrt{2}$ of its maximum value at frequencies in the cut-off locus; the modification form is:

$$H(u,v) = exp \ [\ln 1/\sqrt{2}] \ [D_o/D(u,v)]^n$$

b-1. Restoration techniques

These techniques aim to improve a given image in certain ways. Restoration is a process which attempts to reconstruct or recover an image degraded by using some a-priori knowledge of the degradation phenomenon. Early techniques for image restoration were oriented towards modelling the degradation and applying the inverse process to recover the original image; they were derived mostly from frequency-domain concepts.

b-2. Degradation model

The degradation process can be modeled as an operator H which in the same time with an artificial noise term $\eta(x,y)$ operates on an input image f(x,y) to produce a degraded image g(x,y). The image restoration approach can be interpreted as a method of obtaining an approximation to f(x,y), given g(x,y) and a knowledge of the degradation by the form of H operator. Knowledge concerning $\eta(x,y)$ is mostly limited to information of statistical data (fig. D-1).

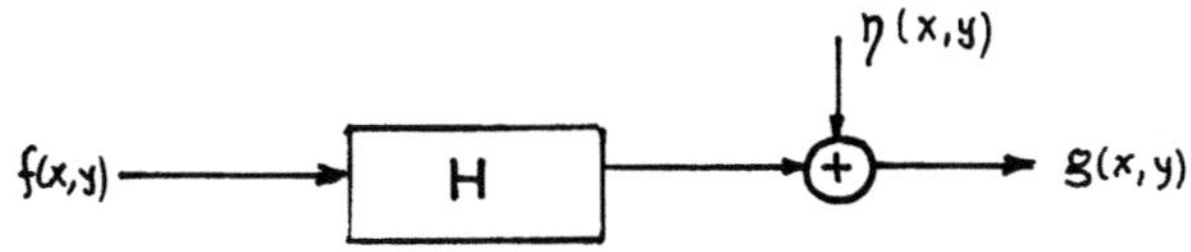

Fig. D-1

The mathematical representation of degradation process is given by:

$$g(x,y) = H\, f(x,y) + \eta(x,y) \quad (1).$$

The operator H is chosen as position or space invariant and with this condition :

$H\, f(x-\alpha, y-\beta) = g\,(x-\alpha, y-\beta)$, is an expression indicating that for any f(x,y) and any α and β the response at any point in the image depends only on the value of the input at that point and not on the position of the point. Many types of degradation can be approximated by linear and position invariant processes.

It is verified that all the terms of Eq. (1) can be expressed in a matrix form as :

$$\mathbf{g} = \mathbf{H\, f} + \mathbf{n}.$$

Finally the entire Eq. (1) is recognized as two-dimensional Fourier transform terms as :

$$G(u,v) = H(u,v)\, F(u,v) + N(u,v) \quad (2)$$

This means that the large system of equations implicit in the model given in Eq. (2) can be reduced to computation of a few discrete Fourier transforms. Therefore, the image restoration approach is commonly referred to as an inverse filter method.

In this lecture it is not appropriate to expose the general solutions to the restoration problem and this chapter is concluded only by indicating the most usual methods such as Wiener filter and constrained least-squares filter.

c. Histogram equalization

This section introduces in the spatial-domain new techniques modifying the gray value distribution of a digital image while carefully preserving edges and contours of shape.

Consider a digital image with n pixels at 2^m graylevels; each pixel can take a randomly continuous value x from the set 2^m. As in a given human population people's size is a randomly continuous variable so image brightness x takes all the values of the interval (a,b). The distribution function F(x) is the probability P when x takes a value smaller than a.

$$P(x < a) = F(a)$$

The probability of the interval (a,b) is given by :

$$P(a, b) = P[x \in (a,b)] = F(b) - F(a)$$

In the interval $\Delta x = (a,b)$ the probability $P(\Delta x)$ is

$$P(\Delta x) = F(x + \Delta x) - F(x)$$

Thus, the derivative $F'(x)$, named $p(x)$, is by definition the probability density function of x (PDF).

The Cumulative Distribution Function (CDF) is the integral value of $p(x)$

$$CDF = \int_o^{xmax} p(x)\,dx$$

where x_{max} is the maximum permitted gray value.
In a digitized image each n pixels takes one random value r_k from k discrete values; PDF is written as:$p_k = n_k / n$ where n_k is the pixel number at the r_k level;
$p_k(r_k)$ is the Discrete Histogram (DH).

In this case CDF is given by H_k which is the integral function of p_k, so

$$CDF = H_k = \sum_{i=o}^{k} n_i/n$$

Example : A digital image I_k with 16 gray levels (fig. D-2)

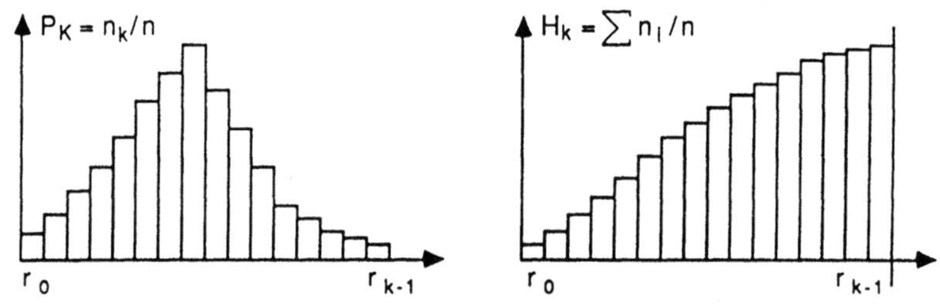

Fig. D-2

By introducing the T transform the input image is modified into the output form : $g(i,j) = T[f(i,j)]$; in this way all the gray levels of the interval (o, x_{max}) are changed by the mapping of PDF and CDF. If x_g is the gray level in g and x_f in f then probabilities of input and output image are respectively $P_1(x_f)$ and $P_2(x_g)$. The probability invariability can be written by the equations :

$$\int_o^{x_{max}} p_1 dx_f = \int_o^{x_{max}} p_2 dx_g = 1$$

The egalization is the process giving the constant value C to the probability density p_2 of the output image such as :

$$\int_o^{x_{max}} p_2 dx_g = \int_o^{x_{max}} C dx = 1$$

After some calculation the T transform is finally given by

$$T(x_f) = x_{max} F_1(x_f)$$

Example : Consider the 64 x 64, 8 gray level image values; the original histogram p_r is shown in fig. D-3.a; s_k values represent $F_1(r_k)$ ans p_s values equivalent to $p_2(x_g)$ provide the resulting histogram in fig. D-3.c.

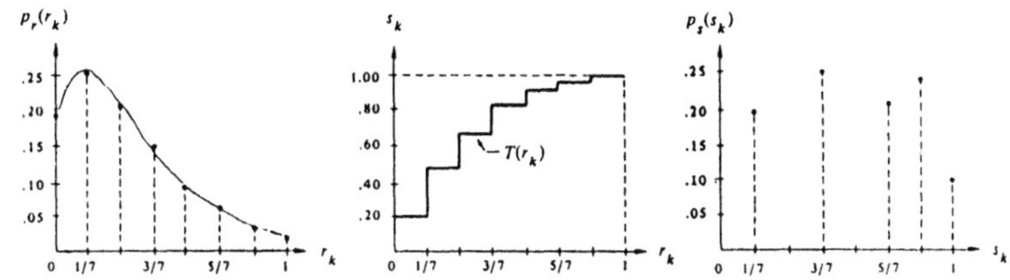

Fig. D-3

d. Hough Transform

The Hough Transform (HT) is conceptually not very different from a match filter; this image pixel processing was indeed suggested to detect lines in noisy point pattern by looking at a parameterized space where evidence can accumulate to indicate the presence of specific line patterns. The axes of the evidence space are known as accumulator arrays and correspond to the feature characterizing the event one wants to detect.

Consider the parametric equation of a line D given in terms of angle θ and distance to origin ρ:

$$\rho = x_i \cos\theta + y_i \sin\theta \qquad (1)$$

In the image plane J, a couple of parameters: (ρ_o,θ_o) corresponds to any point M of a line D (ρ_o,θ_o). In the Hough space H any point of the transform given by (1) is on a sine-curve $S(x_o, y_o)$ whose parameters x_o and y_o are constant (fig. D-4).

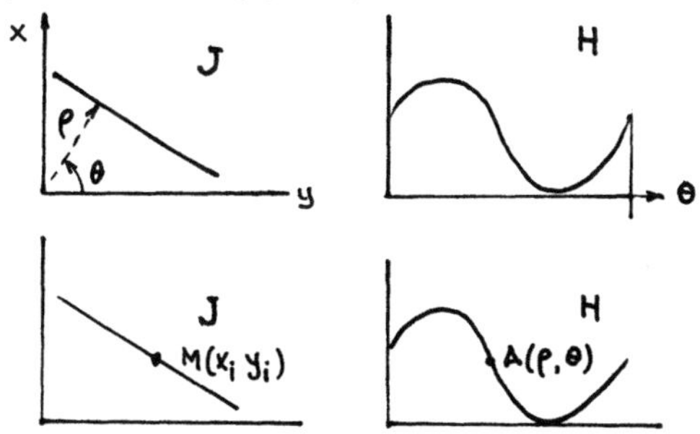

Fig. D-4

The same approach could be extended to the detection of parametric curves as circle and ellipsis. Note that curves of H space corresponding to the points of one line D go through a common point known as the joint vertex.

The practical method of line research consists in quantifizing the H plane according to ρ and θ; for each point (x_i, j_i) of J space the equation f (ρ, θ, x_i, j_i) indicates the cell-values which are entered in the Accumulator-table ACC (ρ, θ); for a given cell a register is incremented every time one point (x_i, j_i) is detected on the line (ρ, θ). When the ACC (ρ_o, θ_o) value is $\geq$k, k points belong to the line (ρ_o, θ_o).

The HT can easily process B & W images containing well or poorly defined objects; but the numerical complexity is high (a combination of 1000 points is higher than 10^8), the accumulator memory must be large and the choice of quantification pitch is a subtle trade-off.

e. Edge detection

By edges we mean the locus of significant graylevel changes; the detection of edges in a digital image is an important first step in image understanding because the boundaries of object in the scene being viewed produce edges in the image ; the edges divide the image into different regions of smooth intensity, leading to a compact description of the scene. The edge detection operation can be carried out in three stages :

- application of an edge operator to the image pixels to produce another image in which the edges form legible structures;
- localization of the edges since the output of the edge operator often produces multiple responses in the vicinity of the edges;
- building up of lists of edges; this process consists of converting the image pixels into a set of lists of consecutive points forming edge segment and a list of connectivity between different edge segments.

Three different algorithms are commonly used for the first two stages : Gradient, Laplacien of Gaussian and CANNY.

e-1. Gradient

Early on in computer vision technology the edge detection was accomplished by forming the gradient of the digital image (ROBERTS gradient). Many operators which approximated the x and y derivatives of an image were proposed; some of the most successful are referred to as Sobel operators; their principle being template (or mask or window) **matching**; the template actually is in general a 3 x 3 array or kernel; the matching procedure is as follows : the centre of the array is moved around the image from pixel to pixel; at every position each pixel is multiplied by the number indicated in the corresponding entry of the array and all the partial results are added in a sum Ps (this is a convolution equivalent product). Ps is then compared to a threshold T; if P>T, the central pixel of the mask is labelled edge.

The two SOBEL convolution operators are defined as :

$$\begin{matrix} -1 & -2 & -1 \\ 0 & 0 & 0 \\ 1 & 2 & 1 \end{matrix} \qquad \begin{matrix} -1 & 0 & 1 \\ -2 & 0 & 2 \\ -1 & 0 & 1 \end{matrix}$$

They are appropriated for detecting edges. The gradient at central point is then :

$$G = [G_x^2 + G_y^2]^{1/2}$$

It is usual to convert the gradient from x and y components to magnitude and orientation.

e - 2 Laplacian of Gaussian (LOG)

The LOG operator detects edges of the zerocrossings of images resulting from the convolution of the original image with a Gaussian smoothing kernel followed by applying the Laplacien - A pixel is a zerocrossing if it satisfies one of the two following conditions :

. The pixel's value is positive and an 8-neighbour value is negative.
. The pixel's value is zero, one 8-neighbour value is positive and the opposite 8-neighbour value is negative.

LOG operator $\nabla^2(G)$ of a digital image f(x,y) can be implemented by first computing the Laplacian and then convolving the result with Gaussian of standard deviation σ; this is a filter operation given as:

$$\nabla^2 (G_\sigma) \otimes f(x,y)$$

Note that a two-dimensional gaussian is

$$G(x) = \frac{1}{\sigma^2} \exp[-(x^2 + y^2)/2\sigma^2]$$

and using finite differences one can write :

$$\frac{\partial}{\partial x} [f(x,y)] = f(x+1,y) - f(x,y)$$

and $$\frac{\partial^2}{\partial x^2} [f(x,y)] = f(x+1,y) - 2f(x,y) + f(x-1,y)$$

The Laplacian operator $\nabla^2 f$ finally takes the form :

$$\nabla^2 f = \frac{\partial^2}{\partial x^2} [f(x,y)] + \frac{\partial^2}{\partial y^2} [f(x,y)]$$

and can be written in a matrix expression (or mask) as :

$$\nabla^2 f = \begin{vmatrix} 0 & -1 & 0 \\ -1 & 4 & -1 \\ 0 & -1 & 0 \end{vmatrix}$$

Finally ∇G^2 is given by :

$$\nabla G^2 = \frac{1}{\sigma^2} [\frac{x^2 + y^2}{\sigma^2} - 2] \exp [-(x^2 + y^2)/2\sigma^2]$$

e-3 Canny edge detector

The present method defines detection and localization criteria as for a class of edges and suggests mathematical forms for these criteria as functionals on the operator response. (ref. 1); this principle points out a single operator shape which is optimal at any scale. The optimal detector has a simple approximate implementation in which edges are marked at their maxima in gradient magnitudes of a Gaussian-smoothed image. The initial image is then convolved with a Gaussian smoothing kernel and the gradient of a Gaussian is formed to provide an image in which edges correspond to maxima (values). The edges are isolated and thinned by non-maximal suppression, i.e., all pixels whose gradient magnitude is not a local maximum in the direction of the gradient are not considered to be edge candidates. Thereafter, the edges are compared to two thresholds; all pixels whose gradient magnitude is above the higher threshold are labelled edges. All pixels whose gradient magnitude is above the lower threshold and which are connected to a pixel whose gradient magnitude is above the upper threshold, are also labelled edges; otherwise, not.

This detector needs a large memory computer and on a practical level, some simplifications are necessary to achieve the computational operations.

In general, depending on the goals they are looking for, the users of different detectors make choices and trade-offs according to the computer capacity they have at their disposal.

2. Segmentation, Contour and Skeleton

2.1. Segmentation

The purpose of segmentation is to partition the image space into meaningful regions; for example, in an image derived from viewing a 3D-scene, the goal of segmentation is to identify regions corresponding to objects in the scene; in air-reconnaissance applications the practical objective is to extract regions corresponding to industrial, agricultural, residential or natural areas. A first category of segmentation deals with methods which are based on examining an image on a pixel-by-pixel basis; a second category deals with techniques which utilize the image information in a prescribed environment. In both cases we may view the segmentation as a decision-making or pattern recognition process whose target is to establish boundaries between regions.

Point-dependent techniques : gray-level thresholding

Image segmentation can be achieved by dividing the gray scale into bands and using thresholds to determine regions. For obtaining these results consider the image histogram; very often a large number of pixels in f(x,y) are dark with the remaining pixels being distributed in the remaining portion of the gray scale. To outline the boundary between objects and the background, the histogram is divided into two bands separated by a threshold T. As the image is scanned, a change in gray-level from one band to the other denotes the presence of a boundary.

Region dependent techniques

In a digital image an edge may be defined as a difference in image characteristics within a local region; a line may be considered as a pair of edges of finite width with a common characteristic in the region between them. These observations indicate that segmentation can be achieved by detecting transitions between regions. The concept of mask matching has found wide acceptance for giving solutions to the segmentation problem.

In the previous section concerning edge detection we already considered amongst several methods, the gradient operator which is represented by the SOBEL operators; but there exists many other operators such as the linear masks mostly dedicated to the detection of horizontal, vertical and oblique lines, as the family of non-linear masks. The last category presents the following models :

Ref. 1 : J. CANNY; A computational Approach to Edge Detection, IEEE Trans. on PAMI; Vol. 8, N°6, NOV 1986, pp. 679-698

KIRSCH
PREWITT
Circular Gradient (FREEMAN method)
VICOM System.

A large number of efficient and devoted mask-algorithms are currently available on the market of the Computer Vision.

2.2. Contour and Skeleton

In this section B and W or binary digital images are only considered and the implementation of contour detection and skeleton research is carried out with binary images.

Edges are constituted by clusters of black points on white background; the segmentation is not achieved at that stage and must be followed by operations such as opening and closing of binary image. The problem is to know how to construct a single closed contour representing a single object; a closed contour corresponds to a path where all the pixels must be connected.
The skeletization is a thinning algorithm; it is necessary sometimes to thin an edge before transforming it into a closed contour; in the same way in order to perform a very precise identification of an object the skeletization allows that object's topological attributes to be well defined; the figure D-4 presents three typical examples of skeleton which constitutes the attributes of the original images.

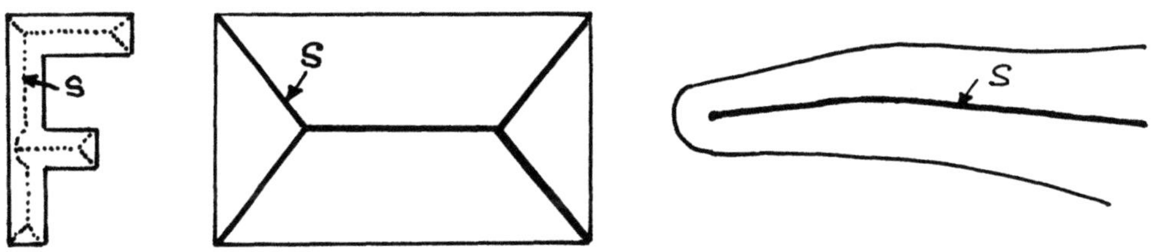

Fig. D-4 : Bold lines and dot are skeletons

E. INVARIANCE, ARRANGEMENT and PATTERN RECOGNITION

1. Invariance

The complexity of computer vision lies in the inherent variability of the input image; such variability can be the result of noise, missing data, topological changes and geometric distortions often due to the information (signal) or transmission (i.e., satellite images). To cope with variability and random changes in the input and to prepare an easy object recognition, the image invariants must be defined and set aside.

a. Contour descriptor

In this section we examine some tools for describing objects such as feature or attribute extractors. In the family of contour descriptors there is the chain-code which is well adapted to the description of line segment with given length and orientation (fig. E-1).

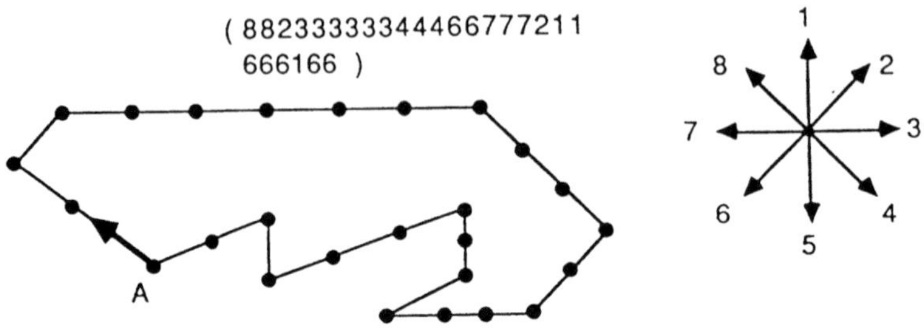

Fig. E-1 : The **chain-code** is given by (88233333344466777211666166)

Thus, there is the shape-number of a given contour which is defined by the smallest first-difference in amplitude; the first-difference of a chain-code depends on the point of origin. The third relevant contour descriptor is the Fourier one, based on the chain rule of Fourier series; the curvature of the object contour is described in the complex space; in a more direct way, if $z = x + iy$ represents the digitized contour (N points) the function z takes N values; the Fourier transform Z(u) of z takes N values and solely 10 values of Z(u) can characterize a contour and therefore an object. The Fourier transform is invariant in scale, rotation or translation; for example, the θ rotation is obtained by multiplying Z(u) by exp $(i\theta)$.

b. Surface descriptor

Topological properties are useful for global description of regions in the image plane; note that topology is the description of a figure which is unaffected by any deformation without tearing or joining (rubbber sheet transforms are accepted). The first topological descriptor is the number of holes H in the surface. The second is the number of connected components C; both number H and C can be used to define the Euler number E as follows : $E = C - H$ (fig E-2)

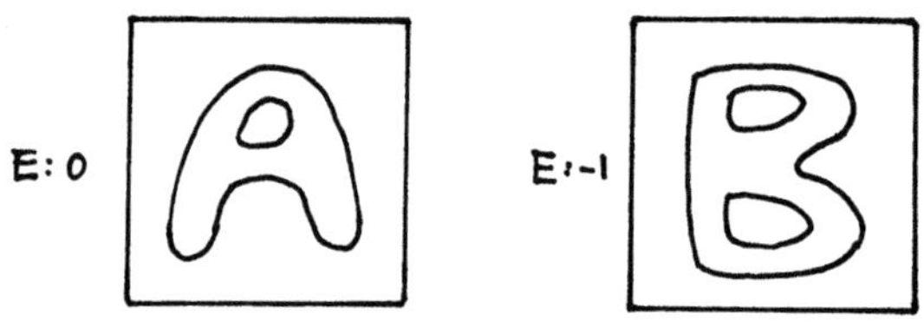

Fig. E-2

Region represented by straight line segments or polygonal networks have a particularly simple interpretation in terms of the Euler number. It is often important to arrange interior regions such as a network into faces and holes. If we denote the number of vertices, edges and faces by V, Q and F respectively, we have the following relationship called the Euler formula :

$$V - Q + F = C - H$$

c. Moments

The geometric invariance discussed so far has been achieved through the use of a distributed architecture. The main alternative is to use moments which are global in nature but unfortunately fault tolerant, yielding poor results for the most part of pattern recognition; nevertheless they are invariant in translation, rotation and scale and their approach is often used for this purpose. Assuming a two-dimensional continuous function f(x,y) we define the moment of order (p+q) by the relation :

$$m_{pq} = \iint_{-\infty}^{+\infty} x^p \, y^q \, f(x,y) \, dx dy \qquad \text{for p, q = 1, 2, ...}$$

The central moments can be expressed as

$$\mu_{pq} = \iint_{-\infty}^{+\infty} (x-\bar{x})^p \, (y-\bar{y})^q \, f(x,y) \, dx dy$$

$$\text{where } \bar{x} = \frac{m_{io}}{m_{oo}} \; ; \; \bar{y} = \frac{m_{oi}}{m_{oo}}$$

and m_{oo} is the total mass function as :

$$m_{oo} = \iint_{-\infty}^{+\infty} f(x,y) \, dx dy$$

A set of normalized central moments can also be defined as follows :

$$\eta_{pq} = \frac{\mu_{pq}}{\mu^{\gamma}_{00}} \qquad \text{with } \gamma = \frac{p+q}{2} + 1$$

The moments can also be used to define a natural and invariant system of coordinates determined by the moments of inertia; the moment of inertia of f(x,y) about the line y = x tan θ crossing the centre of gravity is given by:

$$M_\theta = \sum\sum (x \sin\theta - y \cos\theta)^2 \, f(x,y)$$

When θ equals θ_o (angle for which M_θ is minimal), the line y = x tan θ_o is a principal axis of inertia with respect to f(x,y) and defined as:

$$M_{11} = \sum\sum xy \; f(x,y)$$

$$M_{20} = \sum\sum x^2 \; f(x,y)$$

$$M_{02} = \sum\sum y^2 \; f(x,y)$$

d. Covariance

Statistical invariance is relevant to signal estimation and perceptual modelling of the environment in terms of learning average input image representation. Instead of discussing statistical invariance we shall review here some starting point notions.
Assuming that $X^T = [x_1 , x_2 , ... x_n]$ is a random vector, then the covariance matrix of X is defined as COV(X) :

$$COV(X) = E[(X-\bar{X}) \; (X-\bar{X})^T]$$

where E and $\bar{X}$ stand for expected value operator and mean value of vector X respectively.
Some current applications of covariance matrices are given by the calculation of the so-called Kalman filters and by the estimation of image texture (see further).

e. Texture

Image texture can be described in a variety of ways such as primitives organized by placement rules or as the result of some statistical process. Texture is an invariant and relevant surface characteristic from which basis, shape and motion can be estimated. Much of the work on shape from texture assumes that texture is made up of regular unit primitives; changes in the form such as size, scale, shape and density of such units are then directly related to the layout of the surface. The main approaches for texture segmentation are the statistical and spatial-frequency or spatial/spatial-frequency methods based on convolution transforms. A method used today in texture segmentation is spatial/graylevel dependence described by Haralick (ref. 1); the level dependence estimates the second order joint graylevel distribution for two connected graylevels α and β displaced by the "distance" d (d includes distance and orientation/horizontal). Statistical measurements such as coarseness and entropy can be defined from this and then used to classify/discriminate textures. For example, Haralick's model takes directly into account relative locations and connected pixel graylevels; the graylevels are observed two by two; the pairs thus created are numbered and sorted in graylevel co-occurrence matrices $|\mathbf{C}|$; their terms C_{ij} give an evaluation of the existing point-pair probability satisfying the pair parameters p(d, θ, i, j) where d, θ, i, j are distance, orientation and position respectively.
Other texture descriptors exist corresponding to the following definition:

- *heterogeneity* : $\sum_i \sum_j c^2{}_{ij}$

- *entropy* : $\sum_i \sum_j c_{ij} \log c_{ij}$

- *contrast* : $\sum_{m=0}^{k-1} m^2 \sum_i \sum_j c_{ij}$ with $m = |i-j|$

A large variety of problems encountered within the context of texture analysis would include classification/discrimination, description and segmentation.

Ref. 1 :R.M. HARALICK, K. SHANMUGAN and I. DINSTEIN; "Textural Features for Image Classification"; IEEE Trans. Syst. Man & Cybernetics; vol. SMC-3, n° 6, 1973

f. Labelling

Labelling is the complementary procedure to segmentation; this is based on several operations such as pattern matching and classification (arrangement) of objects in different and distinct classes; the procedure is achieved by symbolic reasoning which consist of manipulating abstract concepts in order to find a reasonable mapping between given image primitives and proposed objects. Here we consider an approach which introduces the notions of probability density map (PDM) and labelling probability matrix (LPM).
Labelling is made up of several steps : standardizing the image position, thus checking the standardized image with respect to the spatial knowledge concerning the object's positions and, finally, making proposals for labelling.
. The first step concerns all the implementations of rotation, translation and scaling based on a crude but robust model of the environment (or world); the world model is a list of spatial relationships which are expressed in rules.
. The second step consists of matching the image segments with a probability density map, i.e. analyzing the probability of each object to be located in any position; this is the learning phase

completed by an overlap operation between segments and a probability density map; the result of the overlap providing the labelling probability matrix.
Labelling presents some interesting properties such as noise resistance due to the comparison between image segments and PDM, absence of combinatorial explosion due to a lack of binary relations. PDM can be much coarser than the original image and consequently contributes to speeding up the matching operation.
In this section we introduce the Black-Board Model (BBM) which is a highly structured DataBase (DB); this DB is operated on by a number of knowledge sources which mainly include the Domain knowledge sources and the Control knowledge sources; all the knowledge sources communicate with each other only through the BB.
The BB system architecture is in general implemented in a system of several (2, 3, 4) Data panels and in one Control panel; the whole system presents four major components i.e. :

- domain blackboard
- control blackboard
- mapping knowledge sources
- image processing procedures library.

The architecture of the blackboard itself is divided into several independent Knowledge Sources (KS) and a Central Data Structure which analyze the current state of the solution search (Fig. E-3)

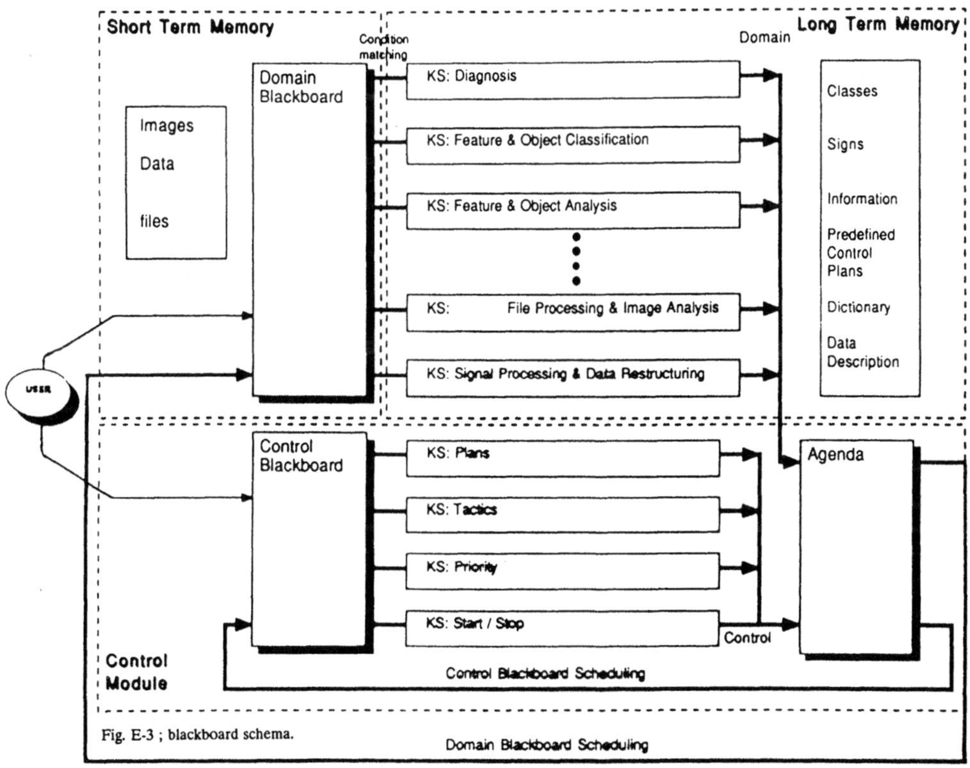

Fig. E-3 ; blackboard schema.

The Blackboard's dynamic behaviour lies in its initialization phase where KS are read into memory, control BB gets its initial plans and Data with associated information which go into Domain BB. Thereafter, the BB cycle can run: the KS try to match their conditions and issue statements which are stored in the Agenda; the Scheduler units interprete the current control plan and estimate the statements again; the highest priority statement is handled by the scheduling Executor taking simple actions and complex tasks into account.

2. Arrangement and Pattern recognition

The main objective of pattern recognition is to classify an object as belonging to one of a number of given classes; linear discriminant or decision functions d(x) are widely used for such a purpose. Consider some object with its set of characteristics x and the direction or weight w; the decision function is given by :

$$d(x) = W^T x + w_0$$

where w_0 is a given threshold.
This expression is equivalent to projecting x on the direction W and comparing the result with the threshold w_0 this indicates the general principles of deterministic decision functions.

2-a. Attribute vector, class, class-vector

The attributes we have already encountered in section E-1 are used to define the attribute-vectors by associating to any object or silhouette S_j a numerical vector **X** with N dimensions; its transposed form is given as :

$$\mathbf{X}^T = (x_1, x_2, \ldots\ldots\ldots x_i, \ldots\ldots x_n)^T$$

where x_i is a component such as Euler number, compacity moment, histogram-moment, texture descriptor, etc.
Learning is the procedure which associates an attribute-vector $\mathbf{X}_j$ to an object S_j.

The class Ω is introduced as an ensemble of similar objects; all the vectors corresponding to the same class of objects are arranged in the form of class-vectors **Y**.
The k-class-vectors are :

$$\mathbf{Y}_m^k = (Y_{1m}^K, Y_{2m}^k, \ldots\ldots, Y_{Nm}^k)$$

where k, m and N are class-index, k-class prototype index and prototype sampling respectively. The class is given by a scalar value : $\Omega = (\Omega_1, \Omega_2, .. \Omega_j, \ldots\ldots \Omega_m)$. When the variety of different vectors X_j is too scattered it is necessary to seek the help of statistical methods; the class element Ω_j is considered as a random variable and its conditional probability density is p (X/Ω_j); the function p evaluates the probability of **X** belonging to the class Ω_j.

2-b. Discrimination Function

Discrimination or decision functions $D_j(\mathbf{X})$ are attributed to any prototype class Ω_j. If **X** is in the class Ω_i then the value $D_i(\mathbf{X})$ is as large as possible. Thus in the attribute-space H_N the partition limit or decision domain between two regions of class Ω; and Ω_j is given by :

$$D_i(\mathbf{X}) - D_j(\mathbf{X}) = 0$$

Minimal distance classification is a method using the point-ensemble distance; there are several ways of implementing the classification : the k-Nearest Neighbour (k - NN) method, the euclidian distance sorting, etc.
For example, consider the Roman alphabet given by the class K among M classes in the space H_N; Y_N^k and **X** are class vector and unknown form vector; the distance D_k of X to K is given by ;

$$D_k = |\mathbf{X} - \mathbf{Y}^k| = [(X - Y^k)^T (X - Y^k)]^{½}$$

All the values D_k are computed and the minimum function D_i is searched for; the form **X** is affected to the class Ω_i whose D_i value is minimum.

When the dispersion of the couples (S_j, X_j) is too large only the statistical method is able to provide reliable results in the classification domain. The most popular method is the BAYES classifier whose role is to assign a random vector X to the class Ω_i. Introducing different definitions such as :

$p(\Omega_i)$ = occurrence probability of class Ω_i
$p(X)$ = occurrence probability of vector X
$p(X/\Omega_i)$ = conditional occurrence probability of X, for X belonging to Ω_i,

we define the likelihood ratio $L_{12}(X)$ as

$$L_{12}(X) = \frac{p(X/\Omega_1)}{p(X/\Omega_2)}$$

and the decision threshold S_{12};

if $L_{12}(X) > S_{12}$ then $X \in \Omega_1$
if $L_{12}(X) < S_{12}$ then $X \in \Omega_2$

F. WORKSPACE REPRESENTATION

a. Robot environment representation

The technology of mobile robotics is continuously developing new concepts in order to sense the environment and to describe it in the most simple and rapid ways. The robot workspace is sensed by several sensory devices; each of them produces intrinsic features of the environment needed for tasks such as location, orientation and navigation. The features, including depth and slant of obstacles (from sonar, laser, u.s. captors) which are in motion or at rest, can be interpreted based on sensor properties and appropriate world modelling. The world model is continuously updated using methods characteristic of spatial reasoning and Data Fusion (DF). DF reduces the uncertainty of interpretation for safe navigation and behaviour.
The problem of environment representation is very important and we give below a simple overview of some of the usual methods.

b. Space configuration method

Location and orientation of an inflexible object can be given by a 6-dimensional vector which is called configuration vector. Say vector A, we note : $C_{espace\,A}$; a configuration is in general formed by 3 location coordinates and 3 Euler angles. The configuration of an industrial mobile robot with 6 d.o.f. is a 6-dimensional vector (ref. 2). On the plane surface when the orientation of a given polygon A is fixed, then $C_{espace\,A}$ is only the (x,y) plane. This property is explained by the location in (x,y) of any reference vertex $\mathbf{rv}_A$; only one vertex is enough to describe the polygon configuration (fig. F-1).

Ref. 2: Tomas LOZANO-PEREZ; "A Configuration Space Approach"; IEEE Trans. Comput., A.C.M. Vol. 20, Feb. 1977.

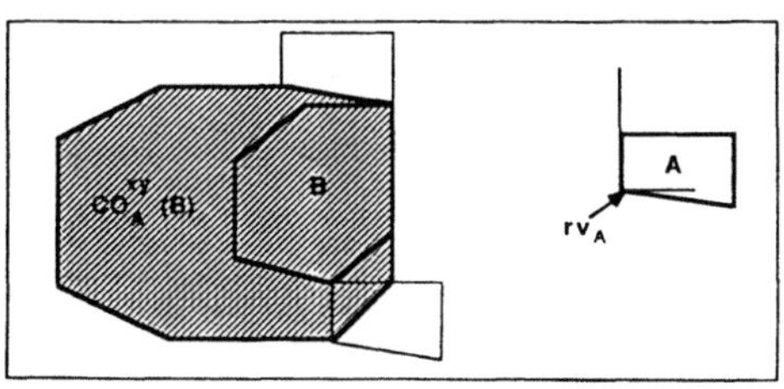

Fig. F-1; polygon configuration

c. VORONOI Diagrams

These diagrams are tools in robot path planning, where the search for a path in an r-dimensional space may be simplified to a search on a r-1 dimensional Voronoi diagram (VD). A diagram VD is based on a measure of distance which is not a true metric. In its simplest form, the measure of distance between a point and a polygon is the maximum of the distances of the point from the half-spaces which pass through faces of the polygon.
More generally, the measure is defined in configuration spaces which represent rotation. The Voronoi diagram defined using this distance measure is no longer a strong deformation of free space, but it has the following useful property : any path through free space which starts and ends on the diagram can be continuously deformed so that it lies entirely on the diagram; (fig F-2) thus it is still complete for motion planning, but it has lower complexity than a diagram based on the euclidean metric.

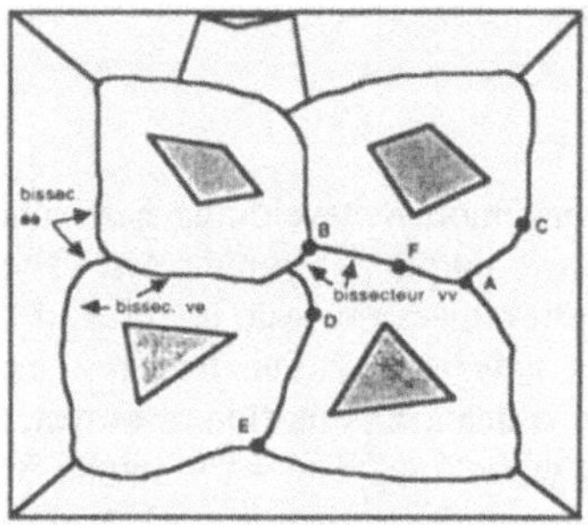

Fig. F-2 The diagram VD divides the space into regions and allows a mobile disc to pass through different obstacles at rest.

d. Generalized Cone

The Generalized Cone principle encompasses the Straight Homogeneous Generalized Cylinder method (SHGC). The apparent contour (AC) of an object forms its limb or external contour; the contour AC constitutes a set of points on the surface of generalized cylinders; conversely the surface generated by AC limits in the space a solid bulk called Generalized Cone or Cylinder. The axis is a free curve of space and the variable plane section is orthogonal to the axis (fig. F-3).

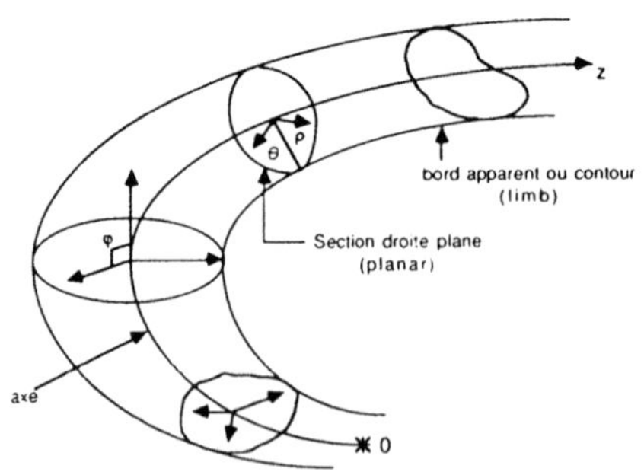

Fig. F-3 Scheme of a Generalized Cylinder

G. ARTIFICIAL NEURAL NETWORKS

1. Introduction and operating system

The Artificial Neural Networks (ANS) also known as Neural Networks (NN) are a suitable approach for Parallel and Distributed Processing (PDP) computation.
Neural networks were originally designed to account for biological memory systems. PDP solutions correspond to a distributed computation, where a large number of highly interconnected simple processing elements operate in parallel. The simple processing elements are simple processors as cells, units or neurons (fig. I-1).

In this section we consider the Feedforward Neural Networks (FNN) built up by grouping a given number N of units in layers; then, the cells of successive subsequent layers are connected in such a way that the output values generated by one cell are amplified and transmitted to the others (fig. I-2).

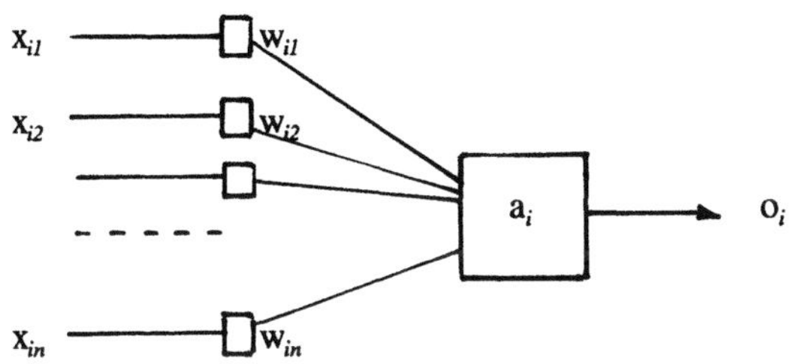

Fig. I-1 : Basic building block of a neural network, also called neuron, unit or simple processor.

At each time t each unit is assigned to receive :
- a column of input values or input vector $\mathbf{x}_i$

$$\mathbf{x}_i = (x_{i1}, x_{i2}, \ldots\ldots\, \mathbf{x}_{in})$$

- a set of weightvalues or weightvectors $\mathbf{w}_i$

$$\mathbf{w}_i = (w_{i1}, w_{i2}, \ldots\ldots\, w_{in}).$$

and to perform operations such as :

- an integration function f_1 such as $a_i = f_1\,(\mathbf{x}_i\,, \mathbf{w}_i)$
- an output value or scalarvalue o_i given by a decision function f_2 such as $o_i = f_2\,(a_i)$.

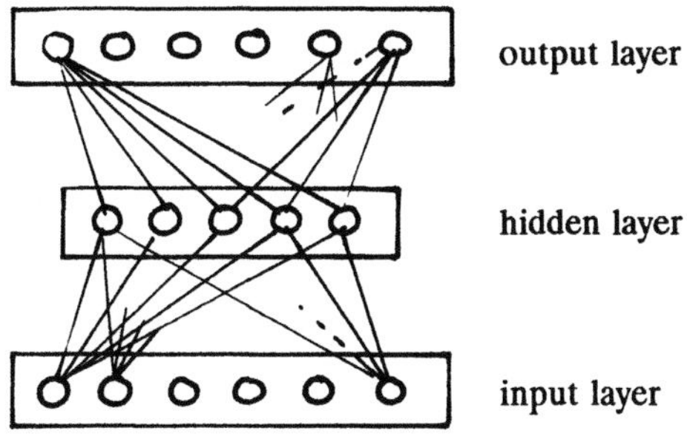

Fig. I-2 : A Feedforward Neural Network is in general composed of 3 layers : input, hidden and output. Each cell of the hidden layer is fully connected to each cell of the input and output layers respectively. Information flows through the FNN from the input to the output layer.

Input information as pixel values of a digital image (input vector clamped to the input layer) is processed locally by each cell in two steps :
- decision or activation levels are carried out by an integration function f_1 computing the inner product of inputvector and weightvector associated to each cell i such as :

$$f_1 : a_i = \sum_{j=1}^{n} w_{ij} . x_{ij} = \mathbf{w}_i . \mathbf{x}_i$$

(see next section below for detailed definitions).

- output levels are generated by a non-linear decision function $f_2 : o_i = f_2\,(a_i)$; this is a very important point permitting in fact the learning process which will be further discussed.
In general the function f_2 is derivable and continuous and has the shape of tanh or sigmoid function:

$$f_2 : f_2\,(x) = k \frac{e^{kx}-1}{e^{kx}+1}$$

With the help of all the weightvectors associated to each cell it is possible to completely define a global network structure by introducing the so-called connectivity matrix :

$$\mathbf{M}_{NN} = (\mathbf{w}_1, \mathbf{w}_2, \ldots\ldots \mathbf{w}_n),$$

where $\mathbf{w}_i = (w_{i1}, w_{i2}, \ldots w_{ij}, \ldots w_{in})$.

A lack of connection between i and j units corresponds to $w_{ij} = 0$.
When the processes are implemented an output information is extracted from the output layer.
Now we can define what is "learning a task"; this is the operation of forcing a given network to map a given set of inputvectors on a given set of desired outputvectors by changing the initial connectivity matrix values.

2. Learning rules

It is easier to use a two-layer network for defining the notion of learning :

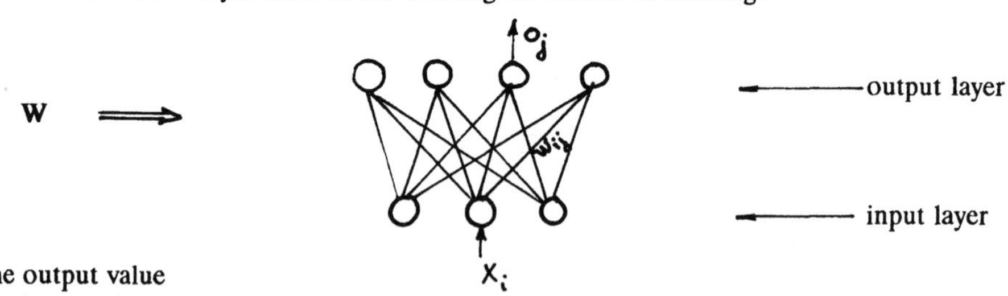

o_j is the output value
x_i is the input value.

In this case, learning is adapting the simplified matrix W for mapping the actual outvalue o_j to a desired one d_j.

In this case, learning is adapting the simplified matrix W for mapping the actual outvalue o_j to a desired one d_j.
The learning procedure is given by the **delta rule**, δ-rule, first historically introduced by Widrow-Hoff in the threshold neuron model (or perception model).

$$\delta\text{-rule} : \Delta_p W_{ji} = \eta \, (d_{pj} - o_{pj}) \, x_{pi}$$

with d_{pj} - o_{pi} = local error at the output unit j

x_{pi} = i-th input value of pattern p

η = learning rate

The supervised learning process consists of minimizing the global error E_T given by :

$$E_T(W) = \sum_{p=1}^{P} \sum_{j=1}^{m} (d_{pj} - o_{pj})^2$$

It is proved that δ-rule minimizes the global error E_T and performs a <u>gradient descent</u> in the error space $E_T(w)$; gradient descent in E_T is equivalent to <u>Gradient Back Propagation</u> (GBP).

This property is extended to multi-layered feedforward neural networks by introducing new expressions for the desired and output values respectively d_{pj} and o_{pj} such as :

$$d_{pj} = \sum_i W_{ji} \, O_{pi}$$

$$o_{pj} = f_j \, (a_{pj})$$

Note here that it is essential for f_j to be continuous and differentiable, in order to write the partial derivative :

$$\frac{\partial o_{pj}}{\partial d_{pj}} = f'(a_{pj})$$

Finally starting from the local errors generated at the output, the δ_{pj} error is computed for all the output units j according to :

$$\delta_{pj} = - \, (d_{pj} - o_{pj}) \; f'(a_{pj})$$

These values are back-propagated layer after layer and the local errors δ_{pj} of all the other units are recursively calculated :

$$\delta_{pj} = -(\sum \delta_{pk} \, w_{kj}) \; f'(a_{pj})$$

The weights between two subsequent layers are then systematically changed. This procedure is repeated until the input layer is reached and defines one back-propagation stage. The whole back-propagation process is repeated for all the learning vectors and defines one learning cycle; repeating learning all over again will minimize the global error.

3. Applications in image processing

A non-linear FNN based on GBP can in principle provide solutions to the non-linear classification problems.

. The first current application is the image detection; the purpose of this operation is to extract a label from a given image; conversely a network is supposed to recognize an image if the corresponding cell in the outlayer generates a high activation level.

Example :

A digital image 64 x 64 is clamped to the input layer 64 x 64 of a given network made of one hidden layer 64 x 64 and one label-output layer (cell number is equal to reduced image number). For a subset of 8 images after 30 iterations, the network allocation and sweeping times are respectively 15 and 13 seconds; learning phase is 5 minutes and total learning time is approximately 1/2 hour.

. The second application is image storing; the network is learnt to generate an appropriate image in response to a given label.

. The third application is image processing by networks characterized by the same cells number in both input and output layers. These networks are able to perform noise reduction, data reduction and some reconstruction operations.

In conclusion, despite the complexity of the inputspace, it is not obvious that the learning algorithm is stable; as a matter of fact the gradient descent in the error space requires several local minima and the convergence may be affected. There is no evidence that neural networks can be made invariant in respect to translation, rotation, changes in gray levels and scaling.

With the exception of FNN, some other models of neural network were recently proposed. The HOPFIELD model is an entirely connected network without singular layer; the network learning is a static and incremental process which does not use an active cells connection system whereas the network relaxation is a dynamic process.
The KOHONEN model is mostly defined by the ability to be self-adapting; this property allows it to perform pre-processing or orthogonalization in the patterns recognition questions.

III. Practical Applications

The aim of this concise lecture on robotics and related computer vision is not to develop very detailed descriptions about all the main concepts envisaged in this large field. Regarding robot navigation and object recognition which are the two main topics in the last chapter, we are satisfied to limit ourselves in the following sections to more essential and fundamental notions and consequently the text will not be greatly elaborated; you will only be given titles and subsequent ideas and principles but during the lecture you will be presented with detailed information on overhead transparencies.

A. OBJECT RECOGNITION

Recognition of specific patterns is the ultimate goal of any visual system. Recognize means to classify and recover the original input object.
In the industrial world people are interested in the design of robust and fault-tolerant systems which demonstrate acceptable degradation, i.e. relative non-sensibility to missing or wrong data and to computer memory damage. Then, these recognition systems must display invariance to affine transformation and perspective distortion. The success of reliable object recognition systems is based on the fundamental following principles:

1. 2D and 3D Object modelling and description

- Curve and Surface Modelling

  . Depth map
  . Bezier curve, B-Spline curve, CAD-CAM curve
  . Object perspective description
  . Generalized Cylinder (SHGC)

- Computer Structured Language applied to object representation

  . C++ , LISP, PASCAL
  . Object-oriented language

- Object Invariant research

2. Pattern Recognition

- Deterministic Method

  .Discrimination lengths
  .Moiré patterns

- Statistical Method

  . Bayes Classifier
  . MINIMAX Method
  . Sequential Probability Ratio Test
  . Decision Tree

B. ROBOT NAVIGATION

Mobile robots like human beings need to build and maintain models of their environment; the models known as visual maps enable the robot to perform tasks such as navigation and manipulation.
Many visual maps employed use the projected 2D-image as a global map within an fixed coordinate frame. If the visual map is to be useful, it has with reliability to represent the complexity of the 3D world rather than an impoverished 2D-projected image. A well made visual map would in fact be a collection of local maps and their relationships; as the observer moves, new informations become available and the view of the world can be rightly updated.
The robot vision, in the same way as human vision, must present a modularity aspect; integrating different modules suggests distributed processing. Data fusion is concerned with integrating

information coming from different modules and coodinating their activities.
Navigation usually supposes moving around, tracking objects at rest and planning a safe route to avoid collision. Such activities apply to factory-robots near the assembly lines and conveyor belts, autonomous land vehicles (USA project A.L.V.) or someday for exploring planetary sites (Mars). All the knowledge and operations indispensable for navigation are given in a non-exhaustive list below:

1. World representation

. Configuration space
. Voronoi-like program
. Plane or 2D-image representation by QUATREE
. 3D-space representation by OCTREE
. Blackboard method (model updating)
. Data Fusion

2. Path planning

. Artificial intelligence system
. Path search in tree
. Hierarchical Data Structure for path finding.

REFERENCES

This is a non-exhaustive list of text books and articles edited in the last decade.

. Alomoinos J; Shulman D.; "**Integration of Visual Modules**"
Academic Press, 1988, Boston, Massachusetts.

. Ballard D.H., Brown C., "**Computer Vision**"
Prentice Hall, 1982, Englewood Cliffs, New Jersey.

. Besançon J.E., "**Vision par Ordinateur**"
Eyrolles Editor, 1988, Paris.

. Blake A., Zisserman A., "**Visual Reconstruction**"
MIT Press, Cambridge, Mass.

. Bracewell R., "**The Fourier Transform and its Applications**"
Mc Graw-Hill, 1978, New York, NY.

. Brooks R.A., "**Robots and Biological Systems**"
Springer-Verlag, 1990, New York, NY.

. Castelman K., "**Digital Image Processing**"
Prentice-Hall, 1979, Englewood Cliffs, New Yersey.

. Duda R., Hart P., "**Pattern Classification and Scene Analysis**"
Willey, 1973, New York, NY.

. Fu K.S., Gonzales R., Lee C.S.G, "**Robotics**"

Mc Graw-Hill, 1987, New York, NY

. Grimson W., Lozano-Perez T.; **"Techniques for 3D Machine Perception"**
North Holland, 1986, Amsterdam, Holland.

. Hillis W.D., **"The Connection Machine"**
MIT Press, 1986, Cambridge, Mass.

. Hoffman D.D., Richards W.A.; **"Visual Recognition"**
MIT Press, 1985, Cambridge, Mass.

. Hopfield J.J.; "Neural Networks and Physical Systems with Emergent Collective Computational Alibies"; Proceedings of the National Academy of Sciences, USA; 1982; 79. pp. 2554-2558

. Lowe D.G., **"Perceptual Organisation and Visual Recognition"**
Kluwer, 1985, Norwell, Mass.

. Le Cun Y.; "Une procédure d'apprentissage pour réseau à seuil assymétrique";
Proceedings of Cognitiva 85; 1985; Paris

. Minsky M., Papert S., **"Perceptions"**
MIT Press, 1987, Cambridge, Mass.

. Moravec H.P.; **"Locomotion, Vision and Intelligence"**
The Robotic Institute, 1983, Carnegie-Mellon University.

. Nilsson J.J., **"Artificial Intelligence Principles"**
Tioga, 1980, Palo-Alto, California

. Pavlidis T., **"Structural Pattern Recognition"**
Springer-Verlag, 1977, New York, NY.

. Rich E., **"Artificial Intelligence"**
Mc Graw-Hill, 1989, New York, NY.

. Rumelhart D.E., Hinton G.E., Williams R.J.; "Exploration in the Microstructures of Cognition"; 1986; MIT Press, Cambridge, Mass.

. Samet H., **"The Design and Analysis of Spatial Data Structures"**
Addison-Wesley, 1989, Reading, Mass.

. Wechsler H.; **"Computational Vision"**
Academic Press, 1990, San Diego, California

. Winston P.H., **"Artificial Intelligence"**
Addison-Wesley, 1984, Reading, Mass.

THE NET REMOTE MAINTENANCE PROGRAMME

D. MAISONNIER, T. REEVE
The NET Team
c/o Max-Planck Institut für Plasmaphysik
Boltzmannstraße 2
D-8046 Garching-bei-München

ABSTRACT. The paper gives an overview of Remote Handling work for the maintenance of the NET device. It sets out to define the basic requirements and constraints on Remote Maintenance of the NET Machine and to describe the philosophy of discipline, based on a comprehensive Remote Handling Manual, to be followed to ensure successful interventions. Individual as well as general concepts for particular maintenance scenarios are presented. Specific Research and Development work completed to date or currently in hand is described. The dilemma between physical demonstration of equipment at reduced or full scale and computer simulation is discussed. Of the compromises arising, the most important is the decision to perform fully remotely all maintenance operations foreseen once during initial assembly of the Machine.

1. General

1.1. BASIC REQUIREMENT

Man access to the NET Machine (fig. 1) will not be possible after the introduction of Tritium. Over the course of the Machine's lifetime its various components will become more and more activated [1]. A fundamental requirement for the Machine design and for the definition of its maintenance procedures is therefore that all maintenance tasks be carried out remotely.

The main consequence on the Machine design has been to segment it in such a way as to minimise the number of components to be disassembled for the most frequent and for the most complex maintenance operations. Another consequence of this foundamental requirement has been on the overall plant layout. Great attention has in fact been given to the definition of transportation routes for the movements of activated components so as to ensure their fast and safe handling.

For instance, 4 equatorial ports have been allocated primarily for the access of handling devices inside the vacuum vessel (VV) where the most frequent maintenance operations are expected (e.g. replacement of divertor plates). Similarly, the replacement of one of the 16 VV wedge sectors, each including a toroidal field (TF) coil, is possible without disturbing any of the adjacent sectors and requires the disassembly of the upper poloidal field (PF) coil only.

M. C. Becquet (ed.), Teleoperation: Numerical Simulation and Experimental Validation, 243–258.

1.2. REMOTE MAINTENANCE GUIDELINES

All maintenance operations are set up according to the Remote Handling Manual [2]. This sets out standard components for use in the construction of the machine, e.g. electrical connectors or fasteners, which have been specifically developed to facilitate remote operations. Standard tools and procedures are also defined where practicable. It must be stressed that no component nor system can be installed in an area with restricted man access unless it complies in full with the Remote Handling Manual's recommendations.

Remote Handling operations fall into four categories, the main distinction being between in- or ex-vessel components. These two are then further broken down into scheduled or unscheduled interventions. The emphasis of NET effort to date has been on the maintenance of plasma-facing components, in particular the first wall armour tiles and divertor plates, and on the qualification of standard components. Blanket replacement techniques have also been vigorously pursued.

2. First Assembly

2.1. REQUIREMENTS

No full preassembly is considered so as to minimise the overall construction time, although a partial preassembly at the manufacturer's works is required to minimise risks during machine erection.

Large components are delivered on site as single units, with the exception of the 4 outer PF coils which must be manufactured on site because of transportation difficulties. The site workshop is also required for the VV sectors final baking.

Large components are transported horizontally in the site workshop and preassembly hall so as to limit the height of these buildings, and then rotated to the vertical position in a special area, used subsequently for decontamination, before entering the Crane Hall.

2.2. BASIC DEVICE FIRST ASSEMBLY

The assembly sequence (fig. 2) of the basic device can be summarized as follows [3]. A fixed reference plane is provided from the outset of machine assembly to ease components alignment, both during first assembly and subsequent remote reassembly. Once the foundation works have been completed, the peripheral cryostat supporting structure and the supports for both VV and TF coils are assembled on the embedded plates already positioned in the concrete with the required planarity. A temporary structure, corresponding to the Machine main axis, is installed to identify the reference centreline which will act as temporary reference abutment surface during TF coils assembly. Each TF coil is preassembled with its corresponding VV parallel segment and thermal shield panels prior to its installation in the reactor hall. The TF coils are bolted together and each parallel VV segment is welded to the adjacent one along the inboard side. Each VV wedge segment is then inserted between two parallel segments using a set of push-pull equipment and the welds between VV segments are completed. After installation of the intercoil structure the central temporary structure is removed and the central solenoid installed in its place. The outer and the upper PF coils are lowered into position whilst the lower PF coil, previously stored under the VV, is raised into position. VV leak tests are performed after assembly of the equatorial ports and of the vacuum pumping ducts, which are connected to the VV and to the cryostat.

The overall duration of the basic device first assembly, considering two main cranes as indicated in para. 5 hereafter, is estimated at 25 months [4]. The Machine first assembly is then completed by the installation of the shielding blanket and of the plasma facing components.

To keep the assembly time down to the minimum only the 'last of a kind' of each main operation (e.g. assembly of the last VV parallel and VV wedge segments) will be carried out using remote maintenance devices and procedures. Bearing in mind that adequate training facilities will be provided, some of them even during the design phase, this is deemed sufficient for validation purposes.

3. Maintenance of Plasma Facing Components

3.1. OPERATIONS

Maintenance of plasma facing components (e.g. divertor plates, armour tiles, Radio-Frequency launchers) and other in-vessel operations (e.g. recovery of debris, dust cleaning) will be carried out by specialized and general purpose end-effectors mounted onto in-vessel transporters.

For divertor replacement access must be gained from above or below the Machine to cut the cooling pipework with equipment operating from the outside of the VV. The transporter device must be entered into the VV via an equatorial port, which itself requires the removal of a shielding plug. Once deployed into the relevant sector of the Machine the transporter must extend either up or down by approximatively 5m to approach the divertor plate (weight 1 tonne, millimetric positionning accuracy). The plate is then gripped, detached and manœuvred to a transfer port on the equatorial plane where it is downloaded onto a transfer unit. Replacement follows the same process in reverse, with new cooling pipework welded to each divertor plate.

3.2. IN-VESSEL TRANSPORTERS

Two alternative approaches are being followed for the design of transporter systems. These are an Articulated Boom System (ABS, fig. 3) and an In-Vessel Vehicle System (IVVS, fig. 4). Since the two systems are fundamentally different, each can be expected to perform certain tasks better than the other. It is therefore intended that the two should complement one another. In practice this means that each system will be designed to specialise in those tasks to which it is best suited, although each should be capable of performing all the required tasks within the Vacuum Vessel. The main role of the IVVS is expected to be the scheduled replacement of divertor plates, whilst the ABS would appear to be better suited to the maintenance of the first wall armour tiles or possibly of a single defective divertor. If possible the same end-effectors will be used for both transporters.

A full-size prototype version of the ABS, known as EDITH (fig. 5), is currently under construction and commissioning, including a simple full-size mock-up of a VV sector. In conjunction with this an End Effector Positioning Unit (EEPU) has been conceived and is currently being manufactured. The EEPU is intended to be mounted at the boom tip and to provide the vertical translation movement required to position the various end-effectors, in particular a divertor plate gripper and servomanipulators. Tooling has also been designed for the attachment and removal of first wall armour tiles and is again shortly to progress to the manufacturing stage [5, 6].

Work on the IVVS, which was a later concept, is not as far advanced as that for the ABS. A conceptual design has been produced [7, 8] and it is intended that on a similar basis to EDITH a full size prototype version of the IVVS, to be known as MARION, will be built and tested using the same Machine sector mock-up.

Handling trials on the prototypes will permit the designs to be refined into usable systems and eventually determine which operations should normally be carried out by which system.

3.3. INTERVENTION TIME

An IVVS with four vehicles working simultaneously inside the VV is expected to require approximately one month to replace all 64 divertor plates. To estimate the overall duration of the intervention must be added the time required for Machine shut-down, equatorial plug opening, equatorial plug closure and Machine reconditioning, for an approximate total of 3 months.

For the ABS, assuming the simultaneous work of two booms with a 90º reach, the replacement of all divertor plates is estimated at 3 months. The overall duration of the intervention is therefore approximately 5 months. This difference is due to the number of 'workstations' working simultaneously (2 instead of 4) and to the lower rigidity of the end-effector support structure (cantilever beam instead of beam supported at both ends).

4. Replacement of Blanket Segments

4.1. OPERATION

In order to replace a blanket segment, which may weigh up to 135 tonnes, it is necessary to gain access to the VV via the upper port. To achieve this the shielding slabs must be removed from the Crane Hall floor and stored. It is then possible to attach a maintenance flask to the cryostat roof at the relevant sector position, cut the welded vacuum seals, remove the relevant portion of the cryostat and bring a double-door closure into use so as to ensure an adequate control of the environment during intervention inside the VV volume. It is also possible to consider a temporary sealing bellows between the VV port and the cryostat roof to avoid breaking the secondary vacuum, but considerations about ease of intervention may outweight the advantages of maintaining this vacuum integrity. It is then possible to work on the cooling pipework for all the blanket segments in that sector (fig. 6).

The central outboard segment, upper half (129 t), is removed first by a straight vertical lift. The central outboard segment, lower half (51 t), is removed second also by a straight vertical lift, leaving access to the remaining segments in the sector. Any of the two lateral outboard segments (134 t each) can then be removed. From a purely geometrical point of view the inboard segments (36 t each) can be removed independently of the lateral outboard, but are to be removed according to a certain order because of the 'step' configuration choosen to protect the VV welds from neutron streaming (fig. 7). Practically the removal of an inboard segment can either be straightforward or require the prior removal of the two adjacent ones. Replacement segments are fitted in the reverse order.

4.2. BLANKET REPLACEMENT CONCEPTS

Earlier work on the replacement of blanket segments was based on the old NET 1 Machine configuration [9] where the blanket ports were located obliquely at the top of the Machine sectors. Various containment arrangements were considered and many handling concepts were introduced, each being required to tilt the blanket segment up to 27° to the vertical in order to remove or insert it.

The most practicable of these arrangements uses a frame comprising four vertical legs. A platform is driven vertically on guide rails within each of the legs. Upon the platform a bridge runs on rails and carries a trolley to provide the necessary x-y motion. The trolley is the base for the blanket gripper unit with the required tilt drives. It was also necessary to contain the whole device within a Contained Transfer Unit in order to prevent contamination of the surrounding building. The advent of the present NET 2 Machine configuration has greatly reduced the value of previous work because of the changes in VV upper ports layout and in blanket segments segmentation. These have effectively permitted the dropping of the tilting requirement for the manœuvring of segments, which was very constraining.

Recent work includes the completion of a 1:3 scale mock-up (ROBERTINO, fig. 8) of a torus segment complete with a Blanket Handling Device (ROBERT), based on the concept chosen for the NET 1 configuration but upgraded to the NET 2 configuration, retaining the capability for only a small amount of tilting for fine alignment. This work has proceeded in parallel with the development of a software simulation package which will allow, eventually, both kinematic and dynamic simulations of ROBERTINO and ROBERT [10].

A rigorous analysis with the new configuration has shown that the insertion/withdrawal of an outboard lateral segment requires two main steps: the segment is unhooked using a first gripper from its attachment locks and moved in the horizontal plane so as to allow its gripping, with a second gripper, vertically above its center of gravity prior to its vertical translation through the upper VV port. This sequence is now proposed also for the inboard segments in order to use the same main lifting equipment for all segments. It allows to consider as an attractive alternative the main crane as lifting device for segments insertion/withdrawal.

An investigation has also begun into the integration of the containment and shielding requirements and the machine layout (cryostat roof and upper vessel port) to determine the best procedure for opening up the Machine and attaching the equipment to the port.

4.3. INTERVENTION TIME

Detailed reviews of the logistic requirements and of the time necessary for blanket segments replacement have brought to the fore the complexity of this operation and shown it to be the most constraining from the point of view of Remote Handling [11, 12]. Although there are still various inconsistencies between the overall scheme (i.e. the procedure) and the handling equipment presently considered, it is possible to estimate with a certain confidence the time required for the replacement of a blanket segment. This duration is dependent on which segment is to be replaced because it will require the prior removal of between 0 and 9 segments:

- replacement of a central outboard segment, upper half, will last approximatively 4 months as no other segment is affected;
- replacement of a central outboard segment, lower half, will last approximatively 5 months as the upper half is to be removed first;
- replacement of a lateral outboard segment will last approximatively 6 months as the central outboard segment, upper and lower halfs, are to be removed first;
- replacement of an inboard segment will last either 6 or 11 months, depending on whether it requires the prior removal of the two adjacent segments or not.

4.4 COMPLETE BLANKET CHANGE-OVER

The extended phase of operation planned for the NET Device requires that, before its beginning, the complete shielding blanket be replaced by a driver/breeder blanket. From the feasibility point of view it must be stressed that a complete blanket change-over is not different from a single segment replacement. If we consider however that all segments will be replaced sequencially, the time required to do so becomes prohibitive. It is therefore necessary to consider that work will procede in parallel over as many VV upper ports as possible.

This does not create new problems, but exacerbates the complexity of logistic aspects, the criticality of the Crane Hall layout as far as lifting systems and storage space are concerned (fig. 9) and the need for redundant transportation routes for blanket segments (both 'new' and 'old') [12].

As in the case of an individual segment replacement, it is possible to estimate with some confidence the time required for a complete change-over at 2 calendar years.

5. Ex-Vessel Maintenance

Work on ex-vessel maintenance, except for the development of standards (see para. 6), has been rather limited for two reasons: firstly the machine design was not sufficiently advanced for detailed analyses and, secondly, the conceptual difficulty of developping satisfactory concepts is rather limited. Two aspects have however been looked at in some detail: the replacement of a TF coil and the maintenance of the Torus Vacuum Pumping System (TVPS).

The replacement of a TF coil has been analysed during the definition of the Machine first assembly sequence (see para. 2). The proposed procedure is compatible with remote handling techniques, so that from a conceptual point of view the sequence and the equipment for the replacement of any component of the Basic Device is defined. Components that have been considered in particular are: TF coils (fig. 10), PF coils, VV sectors and VV ducts [3]. The only specific issue worth mention is the preference, in order to reduce intervention time, of two cranes with a capacity of 500 t each rather than a single crane of 900 t [4].

The analysis of the TVPS maintenance [13] has confirmed the limited difficulty of defining ex-vessel maintenance concepts: overhead equipment mounted on cranes is adequate (fig. 11). This does not limit in any way the importance of the effort to be dedicated to ex-vessel maintenance studies during the engineering design stage.

Work to date has confirmed, however, that the most critical issue is the awareness of the system's designer with remote maintenance techniques and his willingness to involve, from the very beginning of his work, a maintenance specialist.

6. Research & Development

6.1. GENERAL

Bearing in mind the stringency of the remote handling requirements for NET maintenance, in particular from the point of view of environmental conditions and also of the very large size and weight of many of the components to be handled, equipment design relies heavily on Research and Development actions to adapt and improve where possible existing technology or, if necessary, to develop new techniques. The fact that maintenance operations must be performed fully remotely, with in most cases not even limited/local man-access, means that we have to be fully prepared from the outset of Machine operation. Final commissioning of equipment may, however, be completed during the assembly of the Machine and the initial non-active period of Machine operation.

6.2. DEVELOPMENT OF STANDARDS AND PROTOTYPES

Much of the conceptual design of Remote Handling equipment has been carried out within national Laboratories (CEA, ENEA, KfK, SCK) and the Joint Research Center in Ispra with a substancial industrial support. In many instances this conceptual design has been backed up with practical work to demonstrate that the principles applied in the equipment are in fact valid, that the equipment is technically feasible and that it is worth being developed further.

Such practical work to date has included the construction of the EDITH and ROBERTINO test rigs described above. Additional rigs have and will be built to assess the performance of various welding, cutting and inspection tools. These are mainly concerned with seal welds required for the integrity of the primary and secondary vacuum, with the assembly/disassembly of the so-called cryogenic connectors and with the assembly/disassembly of more conventional pipework.

Vacuum sealing is realised by two U-shaped lips welded or cut by appropriate tool-heads mounted on a trolley travelling along the lips. In some instances it may occur that two or three lips meet at a 'T' joint, and this situation is also being addressed with the development of specialized tool-heads.

Cryogenic pipework takes the form of concentric tubes (fig. 12 and 13), where Helium flows in opposite directions, which are separated by vacuum. Connectors for such lines must necessarily be bulky and complicated, with all joints being welded to ensure complete sealing. Two such connectors have been designed to make use of orbital welding and cutting tools, and the validity of the design has been demonstrated by trials [14].

Conventional pipework is also being addressed because, due to the very limited space available around the tubes, cutting, welding and inspection will often have to be performed with bore tooling (ie from the 'inside' of the tube) inserted up to 10 metres away from the weld/cut area.

Standard components such as electrical and push-pull fluid connectors and bolted flanges have been defined, based on commercially available items, but modified after experience gained from a series of handling trials using a Master-Slave servomanipulator.

6.3. RADIATION RESISTANCE

The 'Proof of Principle' testing stage which is now in progress employs normal materials and electrical/electronic equipment, which is a reasonable starting point. However, the high levels of radiation which will eventually build up within the Machine will have severe effects on most common materials. It is therefore necessary to develop equipment capable of resisting the environment for a reasonable length of time. The integrated dose which has been assumed for design purposes of the maintenance equipment is $3*10^9$ rad of gamma-irradiation for sensitive components (which must therefore be easy to replace) and at least $3*10^{10}$ rad for the others. The maximum envisaged dose-rate of $3*10^6$ rad/hr [1], which can only be achieved in a small number of existing plants, means that in order to test materials after irradiation to the full dose a campaign lasting more than one year is required, so that work had to begin early in the programme. Work on this programme is well in hand, a first test campaign on motors, resolvers, proximity sensors, cable insulators and glasses being nearly completed [15].

7. Numerical simulation vs. experimental validation

The development of the maintenance scenario and related systems for NET is a long term task of unprecedented magnitude from the Remote Handling standpoint. It relies on a variety of experiences within and without the nuclear field [16], on the pioneering work now in progress at JET [17] and on R&D actions set up to address specific issues (see para. 6).

During the definition and the execution of these R&D activities the choice between the numerical simulation of an operation or its experimental validation with prototypical equipment is often to be made and it is, at times, a difficult one. Because of their long duration, usually several years, the basic question that justified the definition of these activities (what is to be

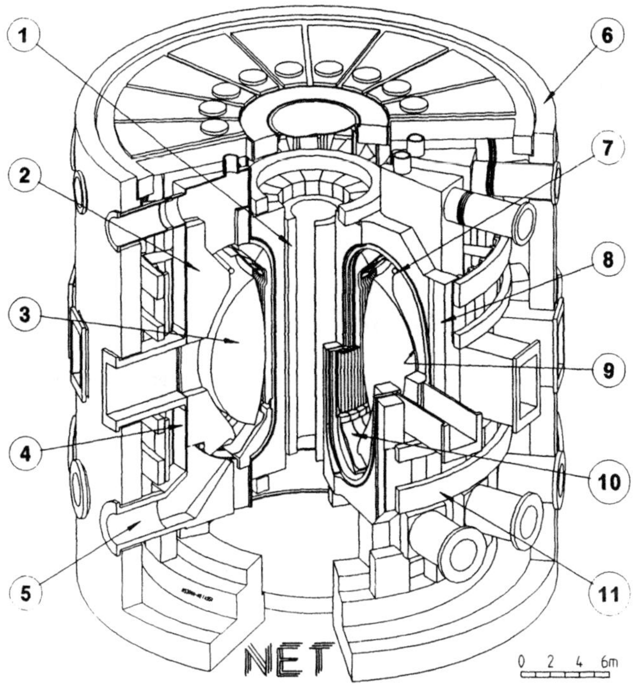

1- CENTRAL SOLENOID
2- SHIELD/BLANKET
3- PLASMA
4- VACUUM VESSEL-SHIELD
5- PLASMA EXHAUST
6- CRYOSTAT
7- ACTIVE CONTROL COILS
8- TOROIDAL FIELD COILS
9- FIRST WALL
10- DIVERTOR PLATES
11- POLOIDAL FIELD COILS

Fig. 1: The NET Machine

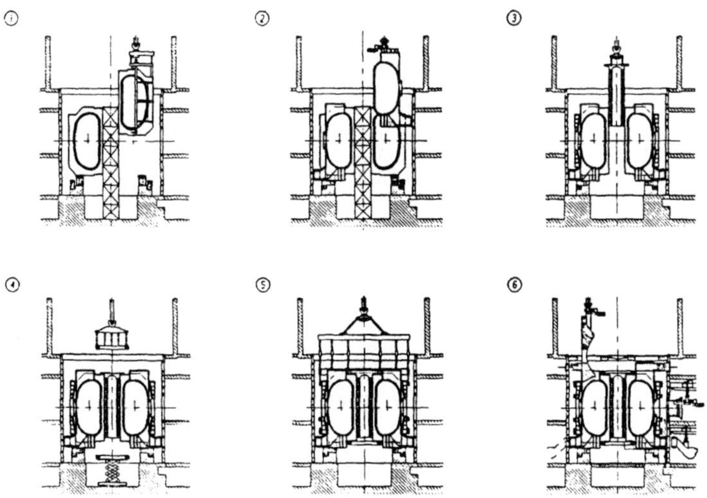

Fig. 2: The assembly sequence for the Basic Device

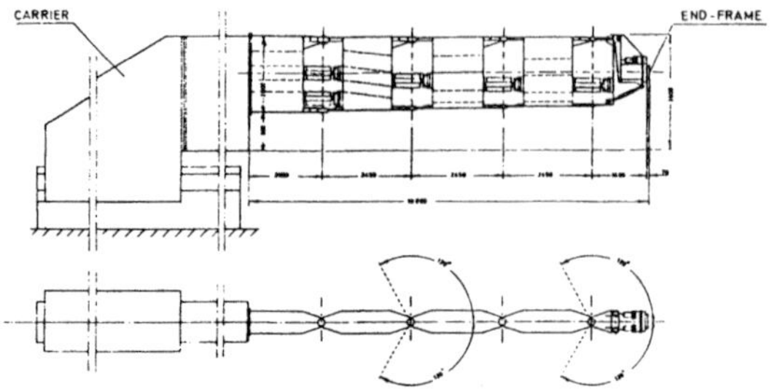

Fig. 3: Layout of the Articulated Boom

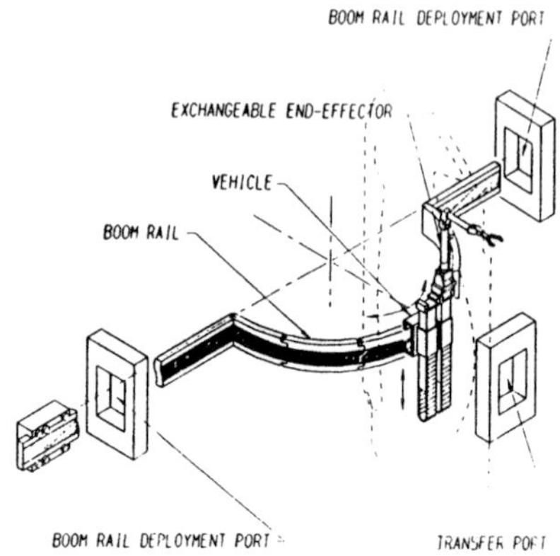

Fig. 4: Conceptual layout of the In-Vessel Vehicle

Fig. 5: The EDITH prototype during commissioning in KfK

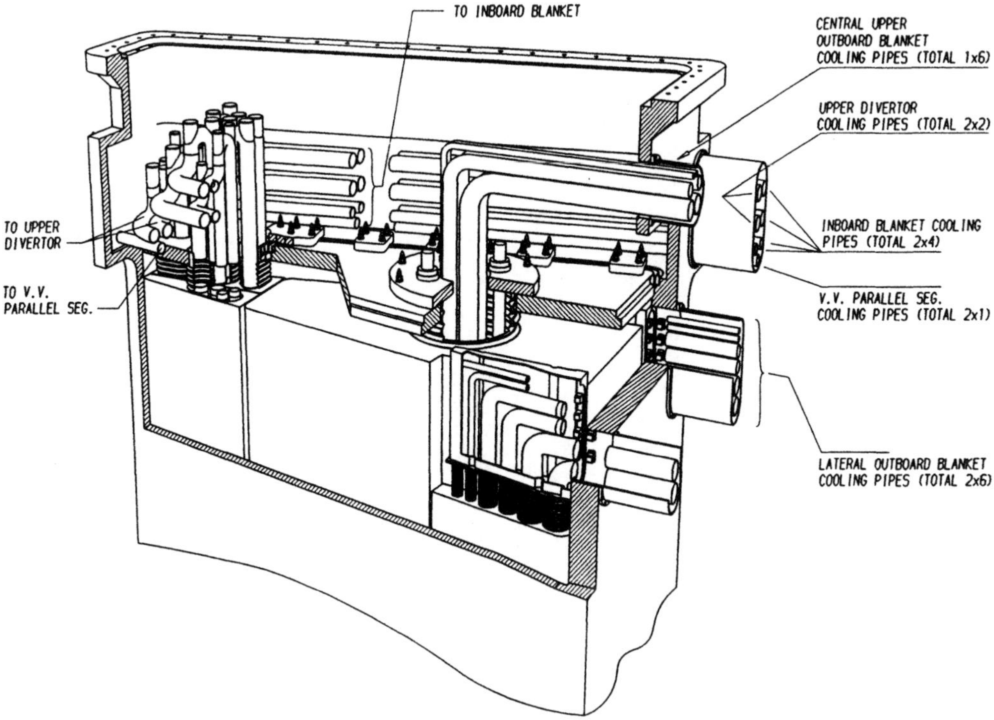

Fig. 6: Layout of the pipework in a Vacuum Vessel upper port

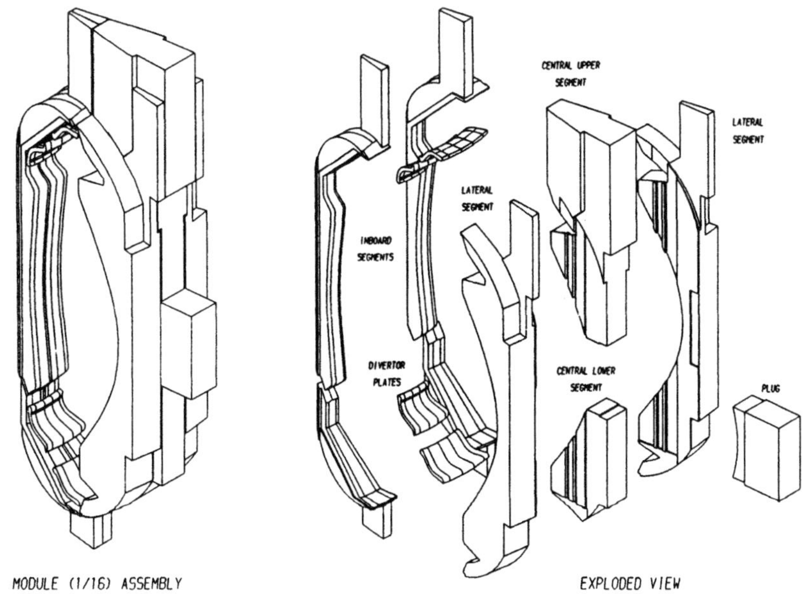

Fig. 7: Exploded view of the Blanket Segments in one Machine sector

Fig. 8: The ROBERTINO test facility in JRC-Ispra

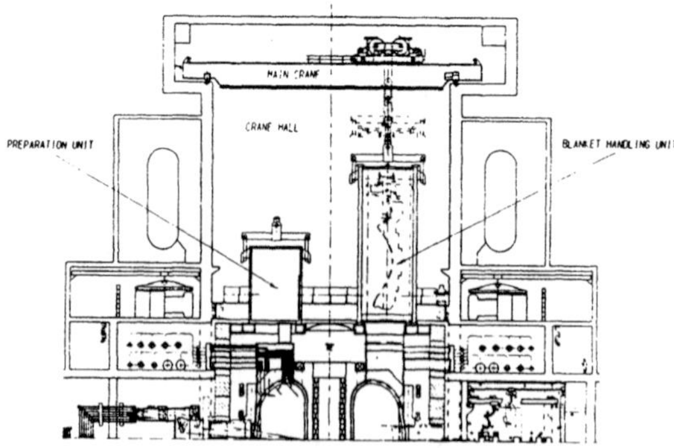

Fig. 9: Blanket handling equipment in the Crane Hall

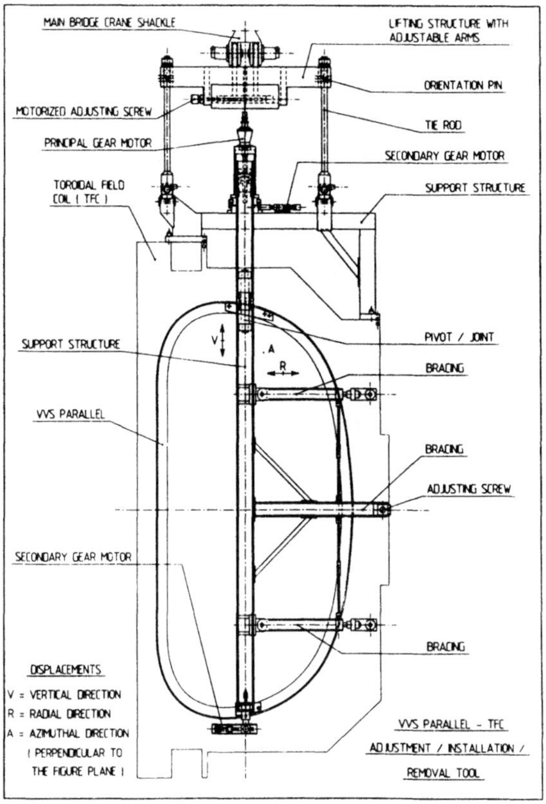

Fig. 10: Special tooling for handling a TF coil together with a VV parallel segment

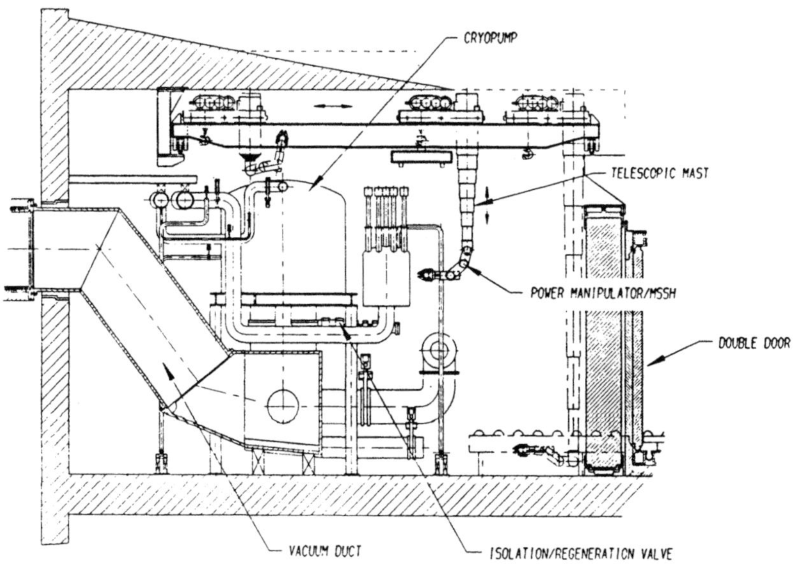

Fig. 11: Handling equipment layout for the Torus Vacuum Pumping System

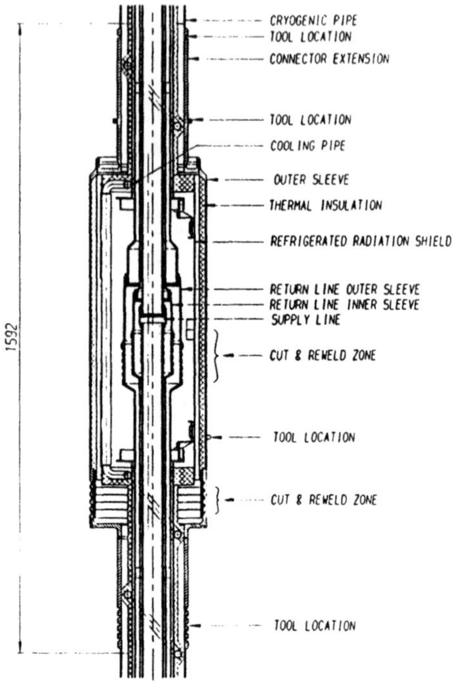

Fig. 12: The 'rigid' cryogenic connector

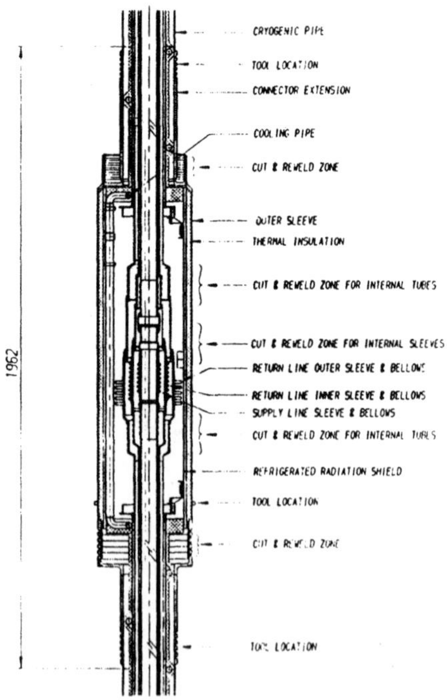

Fig. 13: The 'flexible' cryogenic connector

demonstated and at which stage of the project development) tends to be forgotten and their continuation is then justified with a variety of technico-financial considerations. To demonstrate something when it is not required is a waste of resources whilst to demonstrate it too early, in addition to the commitment of efforts at the wrong time, may preclude the use of technologies or techniques which would have been available at a later date. Considering the rapid evolution of both hardware and software, even a few months difference in the starting date of a specific R&D subtask can have a significant impact.

The capabilities of existing computers and software simulation packages are well suited for the conceptual analysis of maintenance systems with many articulated links or when the environment where the system will evolve is complex. Their 'seductive power' is however a source of disappointment if the limit of what they can achieve is not well perceived. It is worth stressing that the succefull execution of a given task in remote conditions will depend on many details which importance can only be assessed by hardware testing. This testing, to be carried out in parallel with the engineering design, should be limited whenever possible to the 'front-end' of the system (i.e. the end-effector and/or the tooling) so that its implementation does not require the development of a prototype transporter.

The analysis of a complex remote handling task will often require numerical simulation and experimental validation, and the specifications of both activities should be planned together. Care should be exercised when setting the goals to be achieved so that one activity is not asked to provide the answer to a question best addressed by the other. It is also important to review periodically the validity of the technical solution being investigated and of the tools used for its analysis to check wether unforseen technological developments do not offer better alternatives.

Acknowledgements

The authors would like to thank the members of the NET Maintenance Unit and of the various contracting parties (Industries and Laboratories) who contributed to the work described in this paper.

References

[1] C. Ponti: "Activation calculations for NET - Shielding blanket". Technical note ref. JRC-Ispra nº I.88.73, 1988.

[2] NET Remote Handling Manual, C. Holloway editor, 1991.

[3] F. Casci at al.: "The assembly of the NET Machine". 16th SOFT symposium, London (GB), 1990.

[4] F. Olezza et al.: "Assessment of 5 crane systems alternatives and impact on basic device first assembly timescale". Final report for NET contract 90-813, 1991.

[5] A. Suppan et al.: "The NET Articulated Boom: preliminary investigations and justification for a full scale prototype". Report ref. KfK 4809, 1990.

[6] T. Reeve et al.: "The remote maintenance of mechanically attached FW armour tiles". 2nd ISFNT symposium, Karlsruhe (D), 1991.

[7] D. Maisonnier: "Vehicle concept for NET in-vessel inspection and maintenance". IAEA technical committee meeting on robotics and remote maintenance for fusion machines, Karlsruhe (D), 1988.
[8] G. Clement et al.: "A preliminary design of an In-Vessel System for the remote maintenance of NET". 16th SOFT symposium, London (GB), 1990.
[9] Fusion Technology, volume 14 nº1, july 1988.
[10] M. Becquet et al.: "Simulation and experimentation on a 1/3 scale mock-up of the blanket segment removal process". 16th SOFT symposium, London (GB), 1990.
[11] D. Maisonnier: "Analysis of the replacement of the complete shielding blanket by a driver blanket". NET document ref. FTSC-P(91)P-7.5, 1991.
[12] E. de Magny et al.: "Assessment of blanket replacement". Final report for NET contract 91-824, 1991.
[13] C. Holloway et al.: "Remote maintenance of NET/ITER torus vacuum pumping system". 16th SOFT symposium, London (GB), 1990.
[14] C. Barnes et al.: "Remote maintenance of NET cryogenic pipe connectors". 16th SOFT symposium, London (GB), 1990.
[15] V. Massaut et al.: "Gastafiore. Irradiation test of fibre optics and proximity sensors for use in remote handling and maintenance". 16th SOFT symposium, London (GB), 1990.
[16] G. Clement: "Remote handling work relevant for fusion". 2nd ISFNT symposium, Karlsruhe (D), 1991.
[17] T. Raimondi: "The JET remote handling system". Proceedings of this Eurocourse (ref. NSE/91).

COLOR PLATES

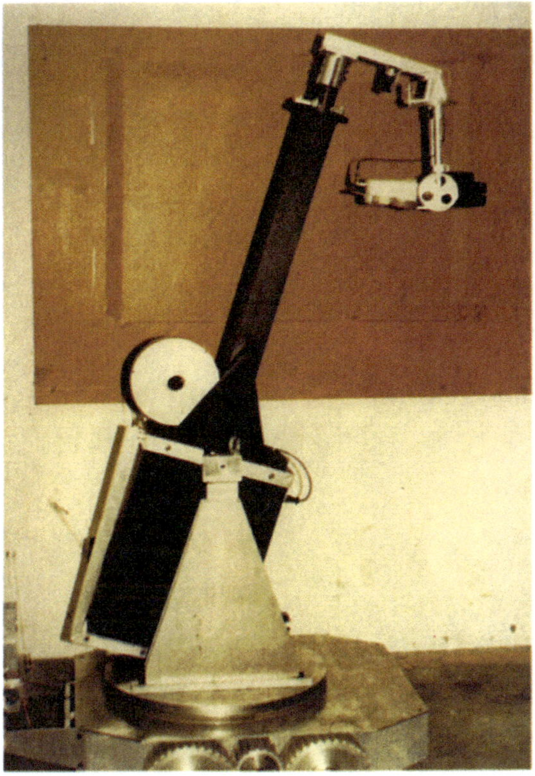

Fig. 1 from page 17.

Fig. 2 from page 17.

Fig. 1 from page 89.

Fig. 3 from page 91.

Fig. 12 from page 106.

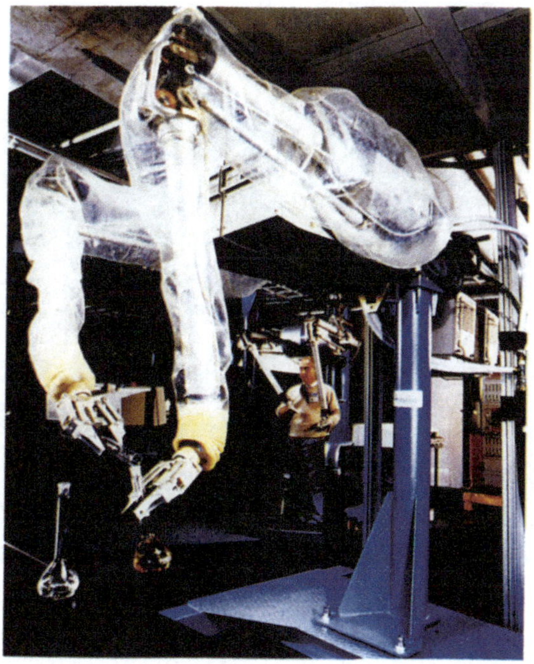

Fig. 1 from page 141.

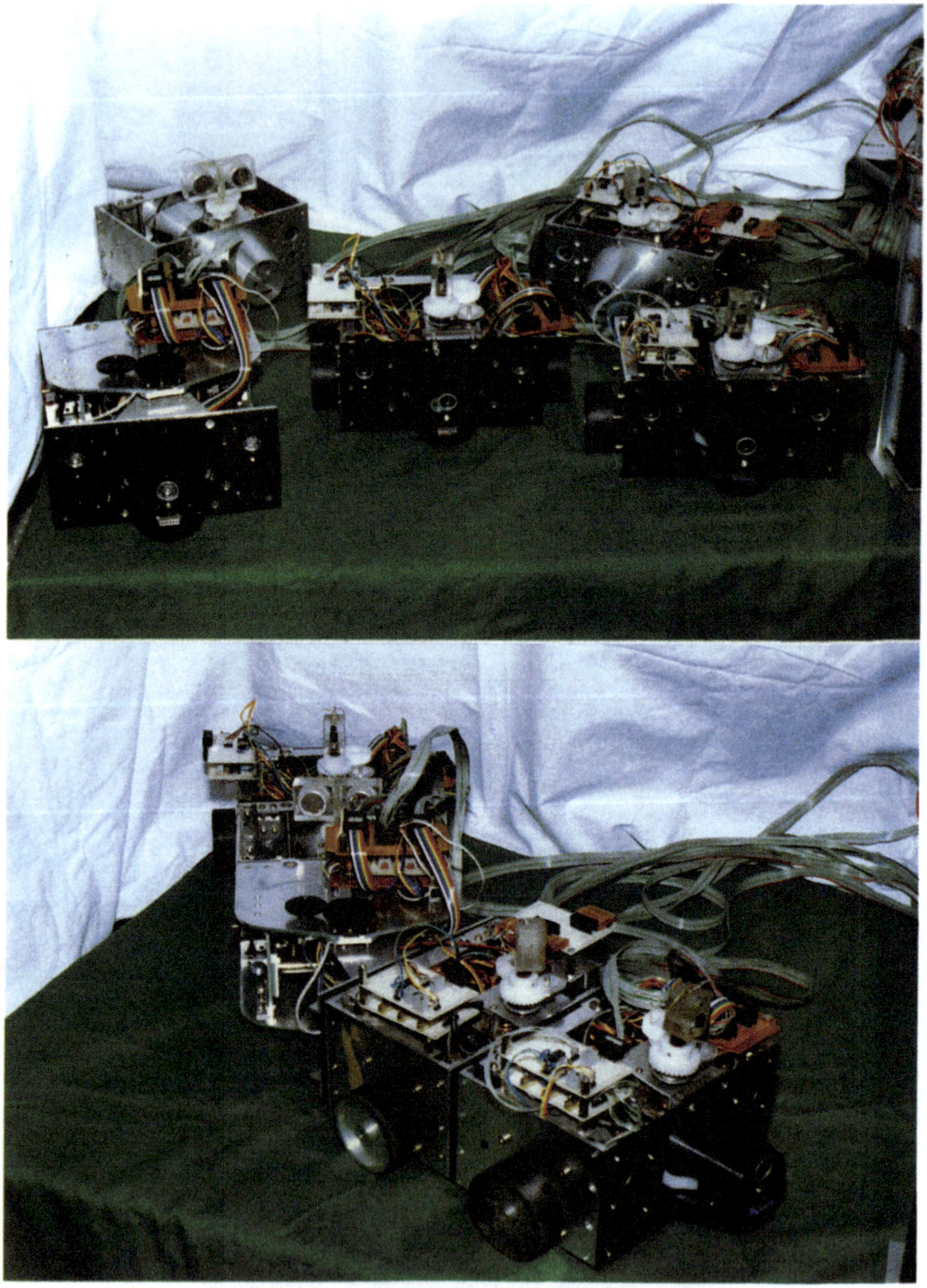

Fig. 8 from page 204.

Fig. 9 from page 205.

Fig. 8 from page 254.

Fig. 5 from page 252.

GPSR Compliance

The European Union's (EU) General Product Safety Regulation (GPSR) is a set of rules that requires consumer products to be safe and our obligations to ensure this.

If you have any concerns about our products, you can contact us on ProductSafety@springernature.com

In case Publisher is established outside the EU, the EU authorized representative is:

Springer Nature Customer Service Center GmbH
Europaplatz 3
69115 Heidelberg, Germany

Batch number: 10607320

Printed by Printforce, the Netherlands